Chemistry's Bridge:
Supramolecular Science

Chopra

Contents

CONTENTS

Epilogue

Outreach

Chapter 1

Introduction

1.1 Supramolecular chemistry

Supramolecular chemistry is a highly interdisciplinary field of science that studies chemical species and complex assemblies that are held together through noncovalent interactions.[1] It can be best described as the science of "Chemistry beyond the molecule".[2] Currently, supramolecular chemistry is located in a multidisciplinary area between the fields of chemistry, biology and physics and has evolved into a major field of scientific research.

In nature, biological structures are often comprised of aggregates and structures held together by weak noncovalent interactions. The dynamic nature of these interactions allows for the complicated processes that occur in living systems. However chemists began to unlock the enormous potential of these interactions only recently. The beginning of supramolecular chemistry can be traced back to study of biological systems by Emil Fischer in 1894 who used the famous lock and key analogy to describe the specific nature of enzyme action.[3, 4] This work brought to the forefront two important principles of molecular recognition and supramolecular function, which would become the main pillars of the new field to be known as supramolecular chemistry.[1]

In 1948 HM Powell published a seminal work on a series he called clathrates which were inclusion complexes by complete encapsulation of a small guest molecule such as methanol or hydrogen sulphide within a cavity formed by a host compound.[5, 6] We now have a supramolecular addition product with no covalent interaction between the so called 'host' and 'guest'. This work was the first step towards the design of host molecules that can selectively include and exclude guest molecules by physical and chemical stimuli. This work, along with the discovery of crown ethers, considered the first artificial molecular receptor by Pederson in 1967 and the study of carcerands by Cram led to the establishment of host guest chemistry.[7, 8] Soon after, the term 'Supramolecular chemistry' was officially coined by Lehn to bring together the various aspects of noncovalent interactions.[9] The Nobel Prize in Chemistry was jointly awarded to Pederson, Cram and Lehn in 1987 for their pioneering work in the establishment of this new field.[10–12] Since then, the field has expanded well beyond host guest complexes and led to advancement in area of self assembly processes which include molecular recognition, devices and molecular machines. In 2016. Jean-Pierre Sauvage, Sir J. Fraser Stoddart and Bernard L. Feringa were awarded the Nobel Prize in Chemistry for their design and production of molecular machines, the second prize to be attributed to the growing field of supramolecular chemistry.[13–15]

1.1.1 Noncovalent interactions and host guest chemistry

The noncovalent interactions leading up to the development of supramolecular assemblies cover a wide range of forces: ion-ion, ion-dipole, dipole-dipole, H-bonding, anion-π, cation-π, n-n, van der Waals and hydrophobic interactions.[1, 16] While the strength of covalent bonds ranges between 150 to $1075\,\mathrm{kJ\,mol^{-1}}$, noncovalent interactions range from 1 to $350\,\mathrm{kJ\,mol^{-1}}$. However, although noncovalent interactions are significantly weaker than covalent bonds, the combination of multiple noncovalent interactions can lead to the formation of a stable supramolecular structure. Table 1.1 shows a summary of common noncovalent interactions along with a few common examples.

Through noncovalent interactions, molecules can spontaneously self-assemble into complex, ordered, multicomponent structures.[18] The information directing the assembly is essentially coded in the molecule in the form of its shape, chemical surface and other factors such as size and conformation. The various steps of self assembly involve reversible dynamic processes. Molecular recognition between two or more molecules plays an especially important role in

Supramolecular Interactions	Directionality	Bond Energies ($\mathrm{kJ\,mol^{-1}}{}^{a}$)	Examples
Covalent bonds	Directional	150-1075	CO_2
Ion-ion	Nondirectional	100-350	NaCl
van der Waals	Nondirectional	<5	Inclusion compounds
Ion-dipole	Sightly directional	5-60	Na^+ crown ether complex
Coordination bonds	Directional	100-300	M-Pyridine
Hydrogen bonds	Directional	4-120	Carboxylic acid
π-π interactions	Directional	2-50	Benzene (edge to face)
Cation-π and anion-π interactions	Directional	5-80	$^+N(CH_4)$. toluene

^{a} Values from reference [17]

Table 1.1: Common supramolecular interactions.

biological systems.

1.2   The cucurbiturils

cucurbit[n]uril (CB[n]) are a prominent family of host macrocycles formed by
the condensation reaction of glycoluril and formaldehyde (Figure 1.1). Although the synthesis of what we now know as the hexameric Cucurbit[6]uril
(CB[6]) was first published in 1905 by Berhand and coworkers, the structure
was not clarified, though its complexation with various metal salts and organic
molecules was studied.[19] This molecule was rediscovered in 1981 when Mock

and coworkers successfully crystallised the product of the reaction and fully characterised its structure.[20] 'Behrand's polymer' was found to be a macrocycle formed of 6 glycoluril units bound by methylene bridges. The molecule was named 'Cucurbituril' due to its resemblance with the pumpkin, the most well known member of the *Cucurbitaceae* family. In the following years, the formation of complexes of CB[6] with cations and interaction of its cavity with alkyl chains was studied.[21]

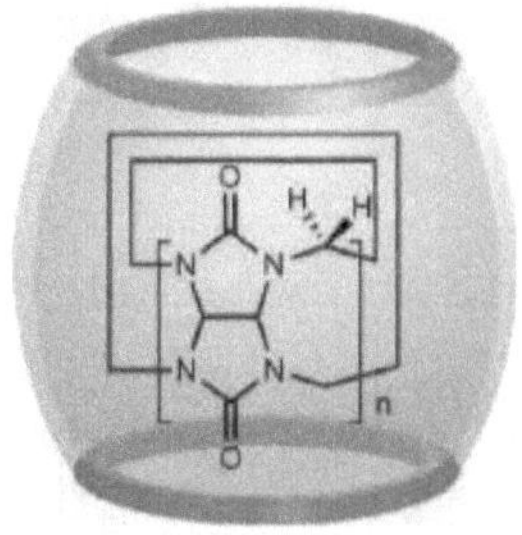

Figure 1.1: Schematic of CB[n].

The research on cucurbiturils fast gained momentum since the early 2000's, which saw the preparation of four new homologous of the CB[n] family by the work of Kim and Day which greatly broadened the scope and applicability of CB[n] chemistry (Figure 1.2). Kim and coworkers in 2000 reported the synthesis, isolation and characterisation of Cucurbit[5]uril (CB[5]), Cucurbit[7]uril (CB[7]) and Cucurbit[8]uril (CB[8]) by varying the initial conditions of Behrand.[22] Soon after Day et al reported a detailed study on the synthesis and reaction mechanism, also isolating the inclusion system of CB[5] into CB[10] (CB[5]@CB[10]).[23] Since then, a number of CB[n] homologues have been discovered, including but not limited to functionalised CB[n], inverted CB[n] and *nor-seco* CB[n].[24] More recently, the twisted CB[14] has been isolated, showing that such larger CB[n]s can actually be formed during the reaction.[25]

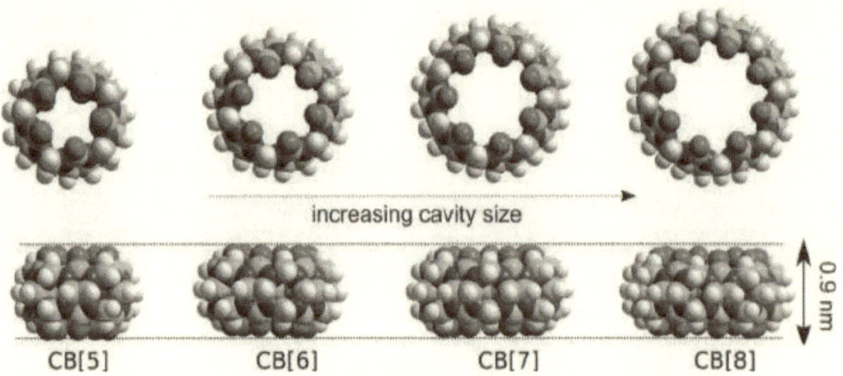

Figure 1.2: The CB[n] family. Adapted from reference [26]

1.2.1 Synthesis of cucurbituril homologues

CB[6] was first synthesised by the acid catalysed condensation of glycoluril and formaldehyde. Kim and day modified the reaction conditions of Behrand carrying out the reaction of glycoluril with formaldehyde in 9 M sulphuric acid at slightly lower temperatures as shown in Scheme 1.1 This resulted in the formation of a family of new CB homologues that could be isolated, the pentamer CB[5], septamer CB[7], octamer CB[8], the decamer Cucurbit[10]uril (CB[10]), and a mixture of non cyclic oligomers.[22]. The typical content of the reaction was 50-60% CB[6], 20-25% CB[7], 10-15% CB[5] and 10-15% of CB[8] and trace amounts of other homologues detected my mass spectroscopy. Higher content of CB[7] and CB[8] can also obtained by synthesis under hydrothermal conditions. The separation of CB[n] homologues is challenging, and is carried out through repeated fractional crystallisation and precipitation based on differential solubility in water, methanol and HCl. The slightly higher water solubility of CB[5] and CB[7] allows their separation from the remaining homologous. which are further separated by solubilising the relatively highly water soluble CB[7] in hot 20% aqueous glycerol solution. The mixture of CB[6] and CB[8] can be further separated either by the dissolution of CB[6] in hot formic acid and recrystallisation in conc HCl.

The mechanism of CB[n] synthesis occurs through the generation of linear oligomeric products followed by their cyclisation. In addition to the work of Day et al, extensive investigation of the reaction mechanism of CB[8] synthesis was performed by Isaacs et al.[27, 28] They performed the isolation and characterisation of several acyclic oligomers and showed that the CB[n] formation follows a step growth polymerisation mechanism, also studying the effect of templation on the formation of CB[n].

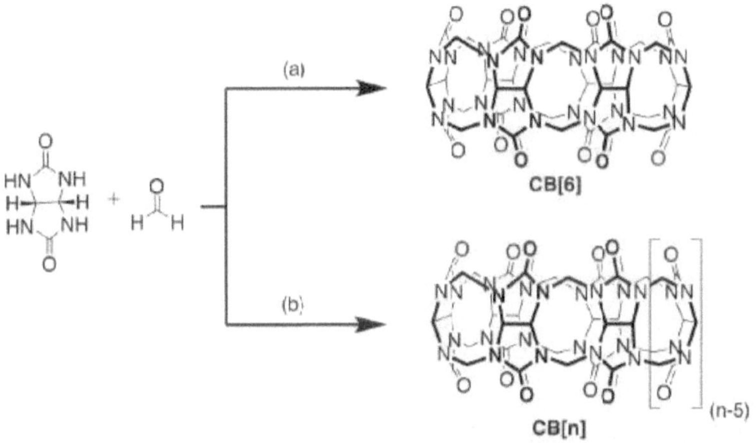

Scheme 1.1: Synthesis of CB[n] from glycoluril under harsh conditions forcing formation of CB[6] and milder conditions leading to formation of a mixture of CB[n] (a)HCl,heat (b)HCl, 100 °C, 18 h.

1.2.2 Structural features and thermodynamics of cucurbit[n]uril guest binding

Structural features

The CB[n] homologues, contrary to other macrocyclic hosts such as calixaranes and cavitands, have a significantly rigid skeleton, clearly shown by X-ray crystallographic studies. It is observed that with the increasing number of repeating units, similar to the cyclodextrin family the height remains more or less unchanged at 9.1 Å. The outer diameter, inner cavity size, portal diameter and volume vary systematically with ring size. Since the portal diameters are 2 Å smaller than the interior cavity, it provides a significant barrier to guest association and dissociation. Table 1.2 shows the structural parameters of the common uncomplexed CB[n] homologues.[29]

Properties	CB[5]	CB[6]	CB[7]	CB[8]	CB[10]
MW	830	99 6	1163	1329	1661
Outer diameter (Å)	13.1	14.4	16.0	17.5	-
Cavity (Å)	4.4	5.8	7.3	8.8	11.7
Inner diameter (Å)	2.4	3.9	5.4	6.9	10.0
Cavity volume (Å^3)	82	164	279	479	870
Solubility in water (mmol dm^{-3})	3	0.05	5	<0.1	< 0.05
Stability (°C)	>420	425	370	>420	-
pK$_a$	-	3.02	2.20	-	-

[1]Values taken from References [22, 23]

Table 1.2: CB[n] structural information

Molecular recognition properties of cucurbit[n]uril

There are two main characteristics that define the molecular recognition properties of CB[n] homologous, that are derived from the arrangement of glycoluril units it its structure. The first and most prominent feature is the urea carbonyl lined portals that lead into the inner cavity. The portals are highly electronegative and this makes them attractive as cation binding sites through the ion dipole effect. In addition, the portals are generally 2 Å narrower than the corresponding cavity. This difference provides a steric barrier for guest association and dissociation and has been reported to increase the persistence of CB[n]-guest complexes.[30, 31]

Though the portal rims of the hosts are highly electronegative, the inner cavities of the CB[n]s are surrounded by the fused glycoluril rings with no accessible functional groups or electron pairs, and are thereby extremely hydrophobic. The inner cavity therefore has a tendency to bind hydrophobic compounds. This explains why some of the highest binding affinity of CB[n]s are towards molecules that contain cationic species connected to alkyl chains

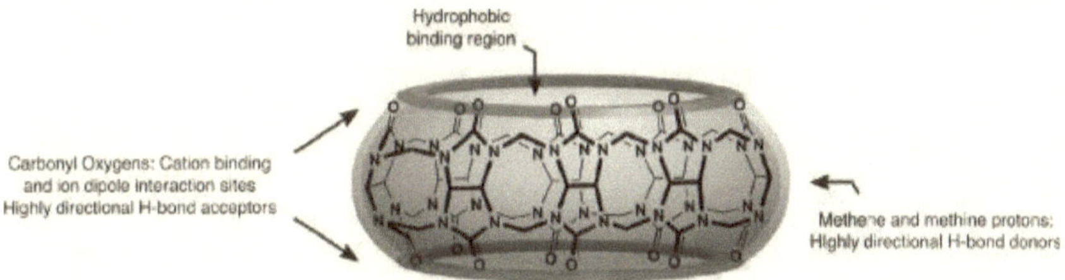

Figure 1.3: The possible interaction sites of CB[n]

or dicationic species that are separated by a hydrophobic core. Such guest molecules can orient themselves such that the cation remains at the portal while the hydrophobic region is within the cavity. Figure 1.3 shows possible binding regions of Cucurbiturils. Another factor to consider in guest binding is size complementarity between the host and guest. It can be seen that the smaller CB[5] due to its small cavity volume best binds gases such as methane, while CB[6] is a good host for alkyl chains. Some of the strongest guests for the larger homologues CB[7] and CB[8] are adamantyl amine, methyl viologen and other small aromatic compounds, while CB[8] is even able to accommodate two guests simultaneously in its cavity.[26, 32]

Non classical hydrophobic effect

The cucurbituril family shows remarkable binding to guest molecules in water. In fact, it was shown that CB[n] shows very high affinities even in the binding of neutral guests which is very difficult to observe in synthetic supramolecular systems. In order to rationalise the the high affinity binding of CB[n] various studies were done to understand the importance of the of size complementary, the hydrophobic effect, or gain in entropy upon binding the release of water, ion-dipole interactions with the rim, and so on. However, in the 1:1 complexation of CB[7] with adamantane derivatives and neutral ferrocenes, ITC data showed that the binding was strongly driven by enthalpy.[33, 34]

This phenomenon can be explained by understanding the contribution of non classical hydrophobic effect. The classical hydrophobic effect is the gain in entropy upon binding by the release of water into the bulk.[16, 35] However, the

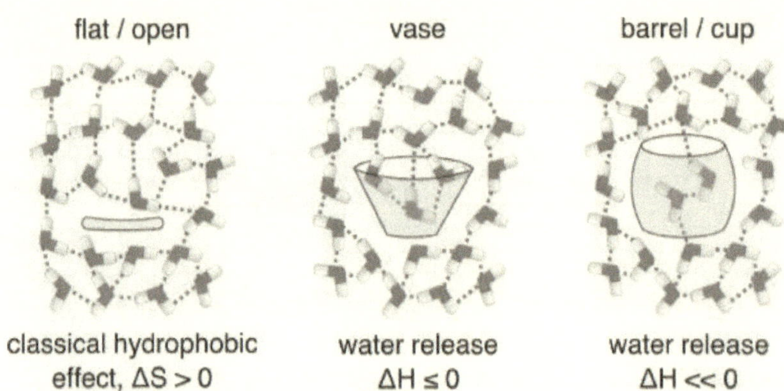

Figure 1.4: Different host geometries and effect of water displacement on guest binding.

non classical hydophobic effect describes the complementary effect whereupon the released water enables better water-water interactions and results in a net gain in enthalpy.[36] In some host geometries, especially in concave hosts, this enthalpic gain can be very high and become a major driving force in guest complexation.[37, 38] As shown in Figure 1.4, certain relatively flat host geometries such as porphyrins barely separate cavity water molecules from the bulk. In these cases the hydrophobic contribution is limited. With the increasing concave nature of the host there is an increase in the seperation from the bulk that can contribute to a higher hydrophobic effect.[39]

The larger cucurbiturils have a cavity volume that is too large to remain empty in solution and always contain water molecules. However, water molecules within the restricted hydrophobic cavity do not have the required number of neighbours to form a stable hydrogen bonding network. This leads to the presence of "high energy" water molecules within the cavity (Figure 1.5). Therefore, restoring these high energy waters into the bulk by displacement upon binding of a guest molecule is strongly driven by enthalpy.[40] De-Simon and coworkers through molecular dynamics simulations and ITC studied this concept in CB[n] homologues in detail and showed that the largest gain in enthalpy during complexation of neutral guests is observed when the largest number of high energy water molecules are displaced from the cavity.[40]

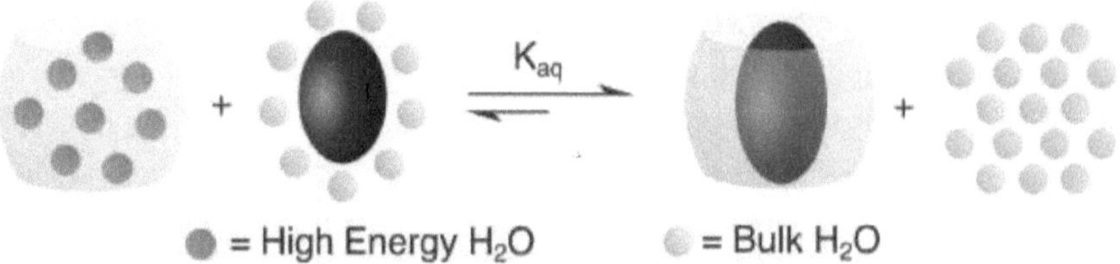

Figure 1.5: Release of high energy water

1.2.3   Cucurbit[8]uril

CB[8] has a larger portal diameter ($6.9\,\text{Å}$) and inner cavity volume ($367\,\text{Å}^3$) than CB[7]. While this results in lower stability of 1:1 host guest complexes with guests also complexed by CB[7], it expands the possibility of additional binding modes as well as stoichiometries for some of these same guests. Similar to the smaller homologous the larger CB[8] does prefer to bind positively charged guests via ion dipole interactions, however in many ways it shows more complex recognition behaviour than just a larger CB[5]-CB[7]. CB[8] easily binds guest molecules that can completely (e.g. adamantyl ammonium) or partially (e.g. dimethyl viologen) fill its cavity.

The most interesting feature of CB[8], however is its ability to simultaneously bind two different guests to form either a 1:2 homoternary complex or 1:1:1 heteroternary complex In the case of heteroternary complexes, CB[8] can first bind an electron poor first guest followed by an electron rich second guest and can serve to stabilise a charge transfer (CT) complex in its cavity, acting therefore as a 'molecular handcuff'. Homo and Hetero complexation offers vast potential in assembly that can be used for switching between favoured and unfavoured complexes that can be achieved through application of external stimulus. They have found a number of applications including drug delivery, advanced supramolecular architectures to name a few.

The study of these homo and hetero ternary complexes in solution can be well characterized by NMR, Mass spectroscopy, ITC and photophysical characterizations. Obtaining single crystal structures of these complexes for X-ray diffraction to understand the packing in solid state proves to be more complicated due to the differential solubility of the host and the complex, with the complex generally showing much higher solubility in water than the CB[8]. There are very limited X-ray crystal structures of CB[8] hetero complexes, with heteroternary complexes being even rarer. The first structure reported is a heteroternary complex of a naphthol and methyl viologen charge transfer complex encapsulated in a CB[8] as shown in Figure 1.6.[21]. Since then ternary complexes of homodimers have been reported.[41–45]

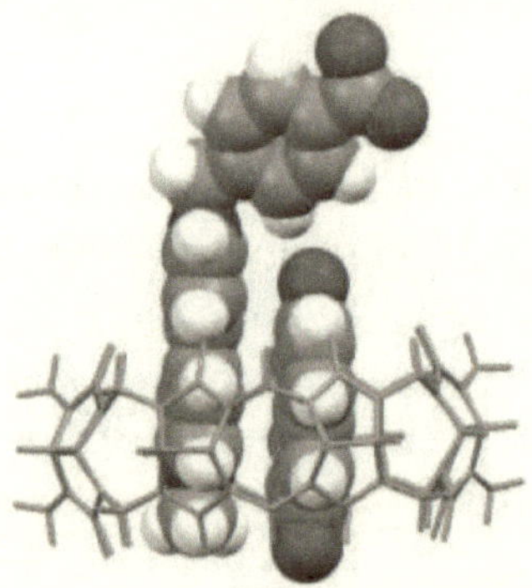

Figure 1.6: Ternary complexes of CB[8]. Adapted from reference [21]

1.2.4 Applications based on cucurbit[n]uril host guest complexation

The remarkable molecular recognition properties of the CB[n] homologues have resulted in numerous applications. This section outlines a few of these applications relevant to the topics described further on in this book.[26]

Supramolecular polymers

Supramolecular polymers are polymers comprised of building block units that are connected by noncovalent interactions.[46, 47] The dynamic nature of the interactions offers great scope for applications in functional materials. Generally, most cucurbituril based supramolecular polymers are based on the CB[8] homologue, due to its ability to bind two guests in its cavity. Supramolecular polymers of CB[n] can essentially be divided into two types, those based on small molecules, or those made from polymer chains or macrocycles mediated by CB[n] binding.

Small molecule based supramolecular polymers: In supramolecular polymers based on small molecules, the CB[8] generally acts as a "molecular handcuff" where it binds together the repeating unit of the polymer. In these cases the concentration of the solution and the binding constant must be sufficiently high in order to obtain a high degree of polymerisation.[48]

Polymer based supramolecular polymers: CB[8] can also bring together high molecular weight polymers and biomolecules by interacting with pendant groups or side chains on the main polymer chain. Scherman and coworkers reported a series of polymers functionalised with terminal electron donor or acceptor groups. Supramolecular diblock polymers were constructed upon encapsulation of the end groups by CB[8].[49] In addition, by randomly functionalising synthetic polymers with these donor and acceptor molecules, in the presence of CB[8] a supramolecular cross linked network was formed as shown in Figure 1.7.

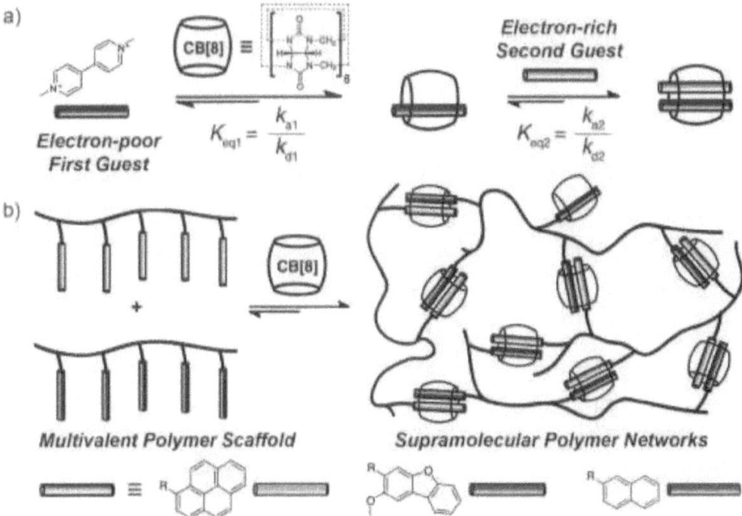

Figure 1.7: Polymers cross linked by CB[8] host guest complexes. Adapted from reference [49]

Encapsulation of fluorescent dyes

It is well known that supramolecular encapsulation of a fluorescent guests can often lead to significant changes in their photophysical properties as a result of the change in its the microenvironment.[50] In context of CB[n], these changes can occur due to the increase in the hydrophobic environment due to inclusion in the cavity, due to restriction of rotation by confinement or even due to the deaggregation of the dye aggregate molecules in solution.[40, 51] In fluorescent dyes which show higher quantum yields in non polar solvents than in water, complexation with CB[n] often increases the brightness of fluorescent dyes by increasing the hydrophobic environment by inclusion in the cavity. Several important classes of dyes with a range of properties show this enhancement in fluorescence including several acridines, xanthenes, quinine- imines, aryl-methanes, coumarines, alkaloids, and many (hetero)aromatic molecules.[52] These complexes have been used to develop biosensors where dye displacement results in a change of fluorescence that can signal the binding of an analyte.

Catalysis

The ability of CB[n] to bring two species in close proximity and in a specific orientation by binding allows it have great potential in catalysing chemical reactions.[24] This was discovered quite early on in 1983 where it was observed that the presence of CB[6] drastically increased the reaction rate of 1,3 cycloaddition reactions of alkynes and alkyl azides. More recently, Nau and coworkers showed that the rate of acid hydrolysis of guests such as N benzoyl cadaverine could be significantly changed by the presence of CB[7].[53] Upon complexation, the reactive group is positioned just at the portal of the cucurbituril which induces the protonation and subsequent hydrolysis of the reaction. The most interest for use as a catalyst remains to be CB[8]. Since it can encapsulate two guests, it can be used as a mini reactor for photodimerisation reactions. It was reported that on encapsulation by CB[8], azastilbene derivatives undergo [2+2] cycloaddition to give stereoselective derivatives on irradiation.[54, 55]

1.3 Strain detection in polymers

Mechanochromic materials are materials that show a change in colour, either reversibly or irreversibly in response to the application of an external mechanical stimulus.[56] Such materials have vast potential applications ranging from the fields of packaging, for example tamper proof seals, to the monitoring of structural integrity of load bearing polymeric materials. The response of a mechanochromic material to strain can be manifested in changes in the absorption or emission properties of the material by mechanically induced modification of either molecular structure, conformation, and or changes in intermolecular interactions.[57, 58] With the increasing use of polymers and polymer composites for structural purposes, the ability to monitor structural health in situ via changes in optical properties is very attractive. The incorporation of mechanophoric systems into synthetic polymers has been carried out through different approaches the most important of which have been listed below.

Direct linking of force activated molecules to polymeric chains

The reversible isomerisation of spiropyrans has been an interesting platform for the development of dynamic materials.[58, 59] Shiraishi and coworkers showed that ring opening isomerisation leading to a colour change could be observed in a spiropyran derivative upon mechanical grinding.[60] Soon after, Moore, Sottos and coworkers first reported mechanochromic behaviour induced by mechanical activation of a reversible ring opening reaction of a spiropyran. Upon activation, the colourless spiropyran was converted to a highly coloured merocyanine thereby acting as a molecular force sensor (Figure 1.8)[61]. Since then spiropyran based mechanophores have been incorporated into a number of soft polymeric matrices such as PDMS for visualising impact strain via monitoring colour change. Further, the reversible colour change in spiropyrans allows the repeated use of these materials for impact evaluation.[62]

The mechanophore was linked to an elastomeric Poly(methyl methacrylate) (PMMA) such that there was a single mechanophore per chain. Upon elongation, a clearly visible colour change was observed indicative of high mechanical stresses.

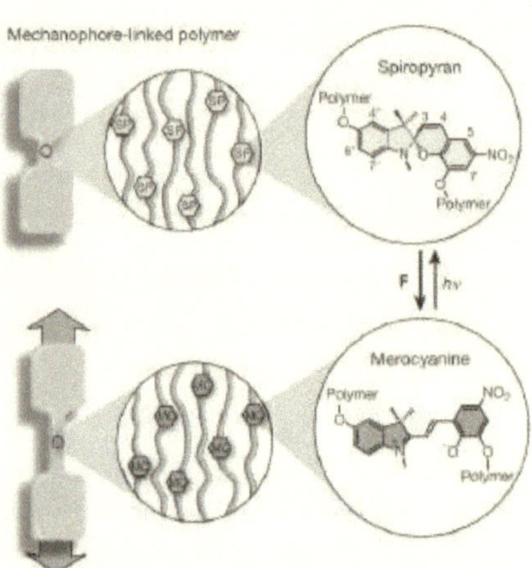

Figure 1.8: Schematic representation of damage reporting by mechanical rupture of covalent bonds. Adapted from reference [61]

AIE dyes and microcapsule approach

Changing the molecular packing and micro-environments of highly conjugated chromophores generally alters the absorption and emission properties quite drastically. In this context, changing the molecular packing of these chromophores through mechanical forces can provide a direct path to the formation of mechanochromic materials.[58] Generally, the aggregation of organic molecules or polymers causes strong fluorescence quenching. However, certain chromophores show increased emission in their aggregated state due to restricted intramolecular rotations and increased electron delocalisation. This effect is called aggregation induced emission (AIE).[56]

Sottos and coworkers used the AIE effect to detect the presence of microcracks in a rigid epoxy matrix.[63] Dilute solutions of an aggregation induced emission (AIE) active chromophore were bound in microcapsules. These microcapsules which were further distributed into the rigid matrix at 10 weight %. Upon rupture of the microcapsules, the loss of solvent resulted in the aggregation of the chromophore and a corresponding appearance of a strong fluorescence along the cracked region. A major advantage of this method was that it could be used to detect microcracks in rigid polymer matrices such as epoxy resins. Weder and coworkers showed that by incorporating a aggregachromic cyano-substituted oligo(p-phenylene vinylene) dye into microcapsules embedded in a polyethylene film, it was possible to obtain a mechanoresponsive material that shows a ratiometric to damage.[64] In these cases, application of mechanical stress by impact, tensile deformation or incision results in a breakage of the capsules, release of the solution and evaporation of the solvent which triggers the aggregachromic response. This approach has great potential especially coatings that allow the detection of barely visible impact damage.

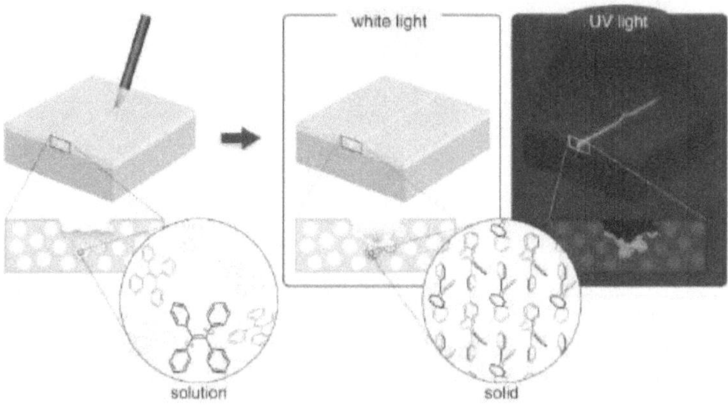

Figure 1.9: Microcapsule based damage reporting polymers

Supramolecular cross linkers

Supramolecular mechanophores are promising candidates for the design of highly sensitive damage reporting systems. Recently, our group reported the use of a tetraphosphonate cavitand based complex as a reporting agent for strain detection in polydimethylsiloxane (PDMS) elastomers.[65] Tetraphosphonate cavitands form highly stable complexes with N-methylpyridinium salts in non polar environments through synergistic cation–dipole interactions between the charged nitrogen and the P=O groups and cation-π interactions between the methyl group and the π-basic cavity. Upon binding of the conjugated guest as show in 1.10, the fluorescence is quenched due to the unavailability of the N-methyl pyridine moiety. The host and guest were incorporated into the

PDMS as random supramolecular crosslinks. The polymer exhibits no fluorescence when unstressed but a clear fluorescence emission of the guest is observed in strained regions when dissociated from the host.

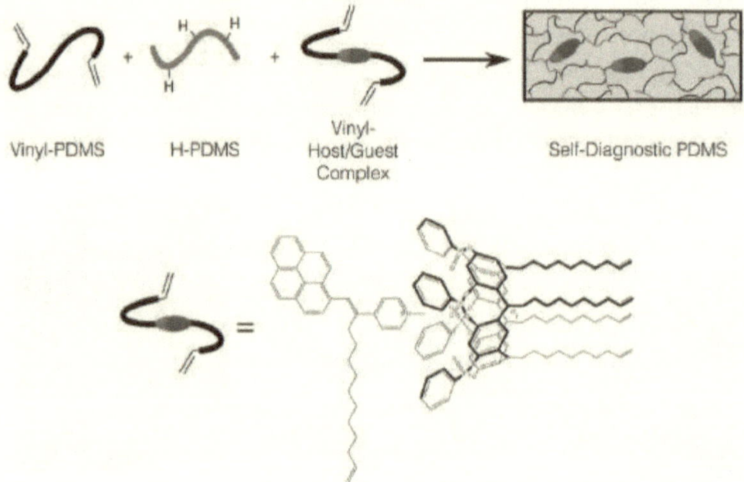

Figure 1.10: Schematic of damage reporting PDMS and host guest complex. Adapted from reference [65]

Recently, the groups of Sagara and Weder reported an elegant approach to mechanically reversible fluorescence in elastomeric polyurethane using the molecular shuttle function of rotaxanes.[66]

1.4   Strain detection in composite materials

The early detection of damage is especially important in fibre reinforced composite materials, since they are high performance materials that are increasingly used for structural applications. Most current approaches to self reporting in composites are based on the inclusion of dye filled capsules, hollow fibres or micro channels that bleed into damaged areas upon rupture and can then be optically detected.[67–70] The inclusion of mechanochromic additives as strain sensors in composites is still less common. Bruns and coworkers showed that enhanced yellow fluorescent protein (eYFP) could be used as a strain sensor to detect damage at the interface of the resin and fibre. eYFP and green fluorescent protein (GFP) are fluorescent proteins that lose their fluorescence when their structure They immobilised the eYFP on the surface of a glass or carbon fibre, and embedded these fibres into epoxy resins to make model composite materials. The specimens were then subject to low velocity impact testing, and it was observed that in damaged regions upon interfacial debonding, the protein is deformed and forced to lose its fluorescence (Figure 1.11).

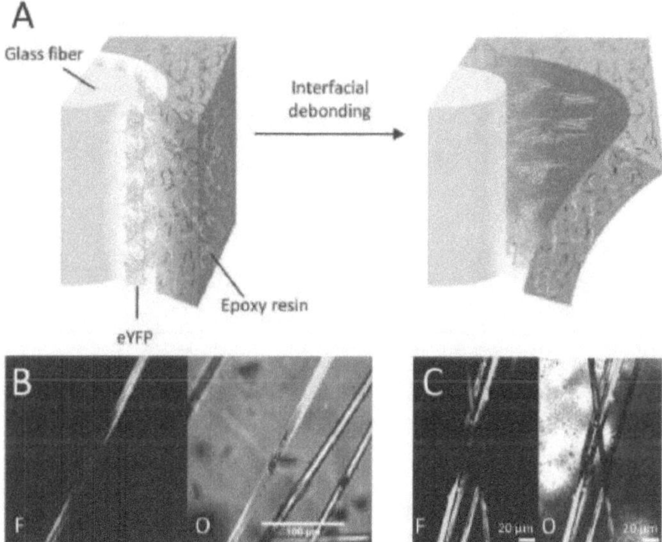

Figure 1.11: (a)Schematic representation of debonding between the epoxy resin and glass fibre, Confocal fluorescence microscopy images in eYFP coated (b)glass fibre and (c)carbon fibre composites after low velocity impact.

Chapter 2

Damage Reporting Carbon Fibre Epoxy Composites

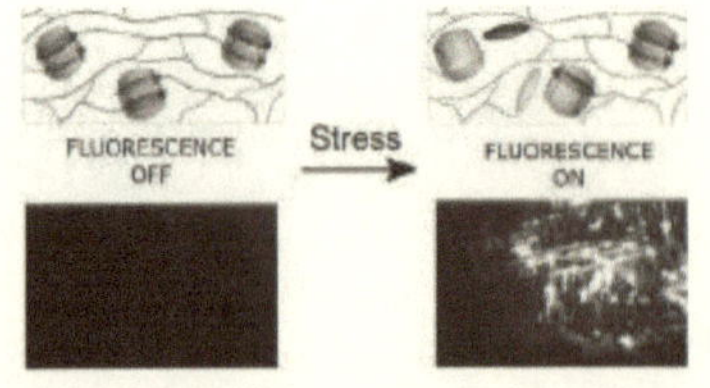

[1]The contents of this chapter were partially published in: A. D. Das, G. Mannoni, A. Fruh, D. Orsi, R. Pinalli and E. Dalcanale, *ACS Appl. Polym. Mater.*, **2019**, 1, 2990–2997.

2.1 Introduction

The development of smart materials that can respond to mechanically induced processes with macroscopic signals has recently gained increasing research interest. Such materials enable the identification of strain prone regions of structural entities which can then be handled before they lead to catastrophic failure. Systems in which the detection of such early stage damage is critical are those where the strength and rigidity of the material is the essence of its function. In the quest to achieve such self diagnostic materials, several groups have employed different strategies. Force-induced chemical reactions can lead to a colour change in mechanophoric molecules such as spiropyrans.[1] These molecules were linked to the polymer chains covalently to create self-reporting materials. In addition, biomimetic approaches to intrinsic reporting of damage were realised in combination with self-healing properties using dye filled microcapsules, hollow glass fibres and microvascular networks.[2–6] These strategies exploited the use of aggregation or separation induced emission of the dye molecules. The majority of this work has been carried out on soft polymeric matrices such as PDMS and polyurethanes. Self-diagnosis in rigid polymers and composites has not been extensively explored. In addition, these methods come with the drawback that relatively large quantities of the active system are needed, which alter the mechanical properties of the polymer and significantly increase the price of the material.

2.2 Carbon fibre epoxy composites

With the rapid development of science and technology there has been an increasing demand for high performance materials with high strength to weight ratio and versatility. Fibre Reinforced Polymer (FRP) composites have rapidly risen to be an important class of materials due to their efficiency as structurally sound lightweight materials. Environmental regulations and rising fuel costs have further driven the use of such composite materials in fields such as the aerospace and automotive industry. More recently, the challenge has been to extend the area of smart materials to composites, giving rise to the development of smart/functional composites.

With the increasing use of FRP composites in fields such as the aerospace and automotive industry where structural integrity and safety is essential, the concept of self-diagnostic polymers that autonomously sense and signal damage is of great relevance. Since carbon fibre reinforced composites are used mainly as structural components, it is critical to recognise any failure or damage at a very early stage to avoid catastrophic failure. Sites of microscopic damage cause the concentration of stress and can further lead to the extension of damage when the material is under load. The sensing of early stage damage via non invasive testing methods could drastically increase safety of these materials as well as their service lifetime. non-destructive testing (NDT) of composites are presently based on X-ray tomography, ultrasound, thermography testing or spectroscopic methods.[7–9] In this context, the preparation of self-diagnostic composites that autonomously sense and signal damage and fatigue is of great relevance for all applications where structural integrity is highly desired.

Current approaches toward self-reporting polymers are mainly focused on the introduction of mechanochromic molecules, excimer-forming dyes or microcapsules. So far, most of these studies have focused on the use of mechanochromic plastics, wherein dye-containing polymers change their colour upon mechanical stress.[10–12] For this, the chromophoric unit, also called the mechanophore, must be appropriately positioned within a polymer chain to experience mechanical perturbation in a controlled manner, and must possess mechanically labile bonds that change through isomerisation or precise bond scission events thus producing distinct optical variations. Mechanoresponsive materials have been realised by directly linking dyes into the polymer chains of elastomeric or on glassy cross-linked polymers. In both cases, the mechanical stress is transduced into the electrocyclic reversible ring opening of the covalently linked dye into the open chromophoric form.[13, 14] Other systems are based on physical effects such as aggregation or separation-induced emission, alteration of the band gap by physical deformation of single walled carbon nanotubes or the force-induced change in photoluminescence of organometallic complexes.[15–17] More recently, a different damage-reporting strategy has been introduced, using aggregation-induced emission of fluorophores confined in core-shell microcapsules.[2, 3] The mechanical damage causes microcapsule rupture and release of the fluorophore and consequent fluorescent emission. This approach allows the detection of scratches and microscopic cracks in stiff polymeric matrices.

Ideally, a reporting strategy for the detection of initial strain is desired to anticipate sudden rupture in composites. Supramolecular mechanophores are promising candidates for highly sensitive self-diagnostic systems, since they are held together by tunable weak interactions. Along this line, the use of host-guest complexes as reporting agents for strain detection in silicon elastomers has been very recently proposed by our group, using cavitands as hosts and N-methyl pyridinium derivatives as guests.[18] The supramolecular cross-linking complex provides a fluorescence response upon dissociation induced by strain well before the formation of cracks. The amount of reporting complex required is so small (1×10^{-6} mol kg^{-1}) that the mechanical and optical characteristics of the matrix are completely preserved. Supramolecular mechanoluminophores have also been introduced in elastomers in the form of rotaxanes. Sagara and Weder used the molecular shuttle function of rotaxanes to mechanically activate reversible turn-on fluorescence in elastomeric polyurethanes under stress.[19]

Translation of these studies to obtain self-diagnostic composites has been limited.[20] The use of a bio-mimetic approach to self-reporting glass and carbon fibre reinforced composites have been reported recently by Bruns and coworkers.[21, 22] The common feature in the two reports is the use of a fluorescent protein immobilised on the fibre surface as a mechanophore to report delamination and fibre fracture. Oxidation and activation of carbon fibre fabric are required for protein immobilisation.

2.3 Design of self-diagnostic matrix and reporting host guest complex

Our approach aims at implementing damage reporting properties into carbon fibre epoxy composites using supramolecular mechanoluminophores transducing the localized stress in the material into a detectable fluorescence signal. The challenge is to produce a fluorescence signal directly linked to the stress driven local breaking of the weak bonds in host guest complexes, leading to the visualisation of emerging mechanical strain in the polymer matrix of the composite. The implementation requires the introduction of a tiny amount of non-emissive host guest complexes in the polymer matrix as supramolecular cross-links, which break apart upon mechanical stress in the strained zone leading to localised fluorescence mission as depicted in Figure 2.1.

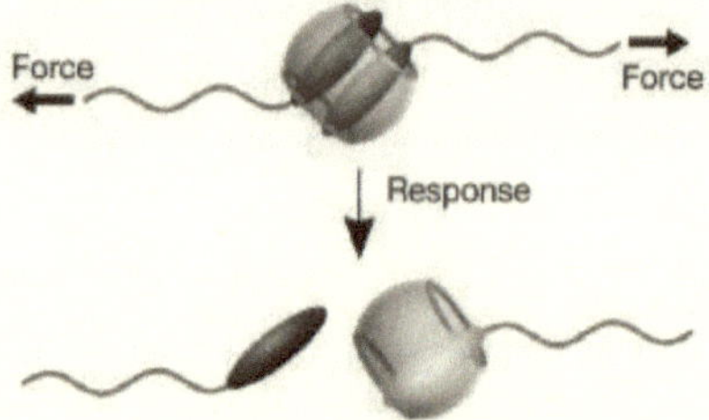

Figure 2.1: Schematic representation of design of damage responsive system

Since crack nucleation often occurs at the surface of structural elements, its detection by optical measurements will allow a full exploitation of the diagnostic characteristic of the proposed materials.

2.3.1 CB[8] ternary complexes as self diagnostic probes

Since CB[8] has the ability to form heteroternary complexes of high stability in polar environments, therefore suitable to be embedded in polar epoxy resin matrices. Moreover, it offers the possibility to select pairs of guests capable of fluorescence turn on and off when dissociated and associated respectively. Working with fluorescence emission in epoxy resins imposes several constraints on the candidate guests. First of all, both the fluorophore and the quencher must be indefinitely stable in the final epoxy matrix and under curing conditions. The choice of a fluorescent probe suitable for use in epoxy matrix required the consideration of additional factors. The absorbance and fluorescence emission of the fluorophore needs to be compatible with the transparency window of the matrix. In recent years majority of commercial epoxy resins based on bisphenol A have been formulated with resistance to UV radiation and have a transmission window above 400 nm. This ruled out the use of more commonly used fluorophores as guests in cucurbiturils, such as pyrene and quinones. We then looked to dyes within the transmission window such as methylene blue, azure A and neutral red. These dyes, however, showed instability to the harsh environment of the epoxy resin.

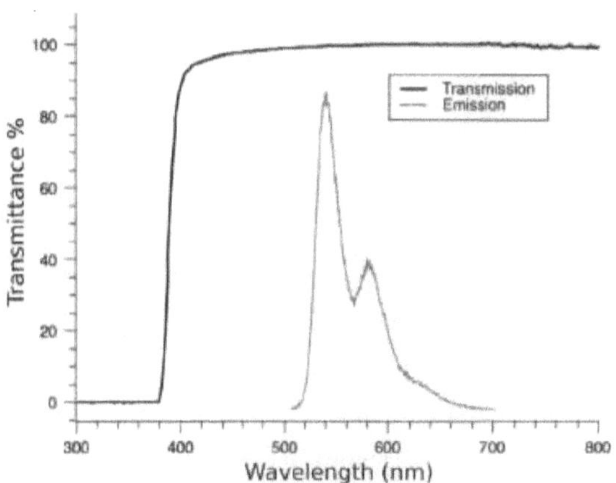

Figure 2.2: Transmission window of Epoxy Resin and fluorescence of Perylene

Highly fluorescent perylene monoimide has all these properties, with absorption maxima above 400 nm, making it an ideal candidate for the fluorescent probe (Figure 2.2). Perylene monoimides and diimides (PDIs and PMIs respectively) are robust photostable dyes and have been widely studied as fluorescence sensors and used extensively as pigments and dyes. However, one drawback is that perylene imides form non fluorescent π-π stacked aggregates in aqueous solution. This has limited their use in biological applications as well as for single molecule spectroscopy. One approach to overcome this issue has been to functionalise the perylene core with bulky or hydrophilic groups. This has indeed allowed PDI and PMIs to be used in a number of applications ranging from fluorescent sensors and labels to molecular to building blocks for molecular wire assemblies. However there is still a significant loss in fluorescence intensity in water unless concentrations are kept very low. It was recently demonstrated that CB[8] can encapsulate PDIs and suitable electron-rich second guests leading to the formation of 1:1:1 ternary complexes that result in the quenching of perylene fluorescence by stabilising a charge transfer pair within the cavity.[23].

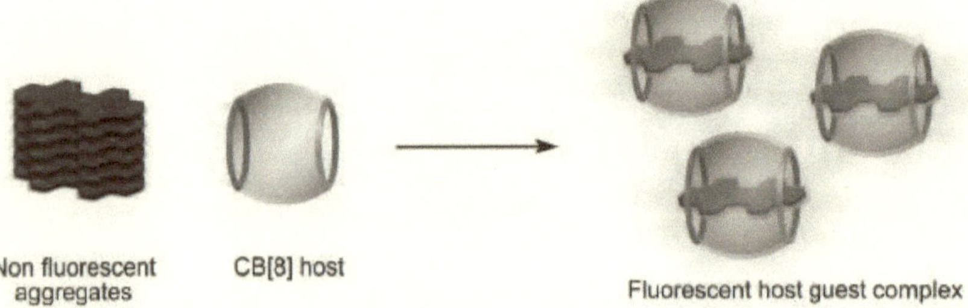

Figure 2.3: Schematic representation of increased fluorescence emission of PDIs and PMIs in water upon complexation with CB[8].

We selected a perylene monoimide (PER) derivative for the present work. Similar to PDIs, perylene monoimide in water has a low fluorescence quantum yield due to its tendency to form aggregates. The fluorescence is dramatically increased by the addition of CB[8].[24] The perylene-CB[8] complex (PMI-CB[8]) has detectable fluorescence even at micromolar concentrations. Two candidate quencher guests were singled out for the purpose from literature: azobenzene (AZO) and dibenzofuran (DBF) derivatives form ternary complexes with perylene imides within CB[8], quenching its fluorescence effectively upon complexation.[25].

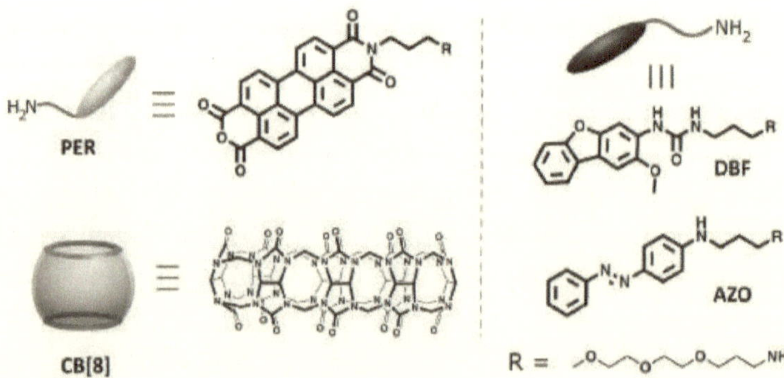

Figure 2.4: Guest molecules synthesised for use in self diagnostic composites.

In order to carry out titration studies, these guests were functionalised with a glycol linker to increase solubility in water. Both fluorophore and quencher guests were also equipped with an amino terminated side chain to participate with the covalent cross-linking of the epoxy resin once added to the curing agent (Figure 2.4)

Initial fluorescence titrations were carried out to study the quenching of fluorescence of PMI-CB[8] on the addition of the third guest. To a solution of PMI-CB[8] in water at [c] = $4\,\mu$M, upon the addition of increasing equivalents of AZO, the fluorescence intensity progressively decreased. Addition of 5 equivalents of AZO resulted in the almost complete quenching of fluorescence of PER as shown in Figure 2.5. This occurs due to the formation of a heteroternary complex inside the hydrophobic cavity of the CB[8], with a face-to-face $\pi - \pi$-stacking of the donor and acceptor.[23, 26]

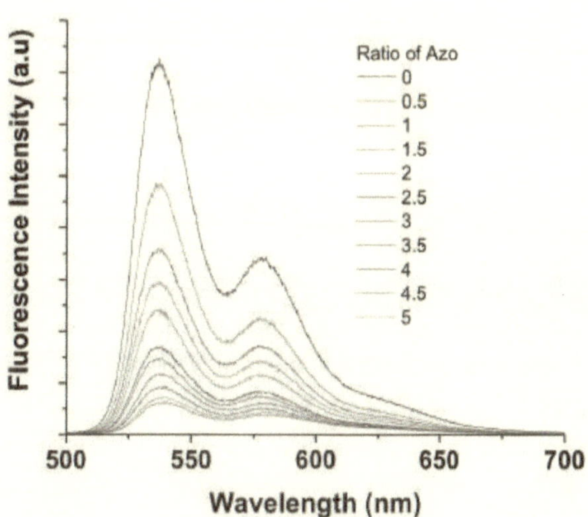

Figure 2.5: Decreasing fluorescence emission of PMI-CB[8] upon the addition of AZO at [c] = $4\,\mu$M in water (λ_{ex} = 480 nm).

DBF behaved in a similar manner and resulted in the quenching of fluorescence of PMI-CB[8] with slightly higher efficiency. As seen in Figure 2.6, the complex shows very low emission upon addition of 2.5 equivalents of the DBF guest. DBF has the additional advantage of showing no absorption in the visible range.

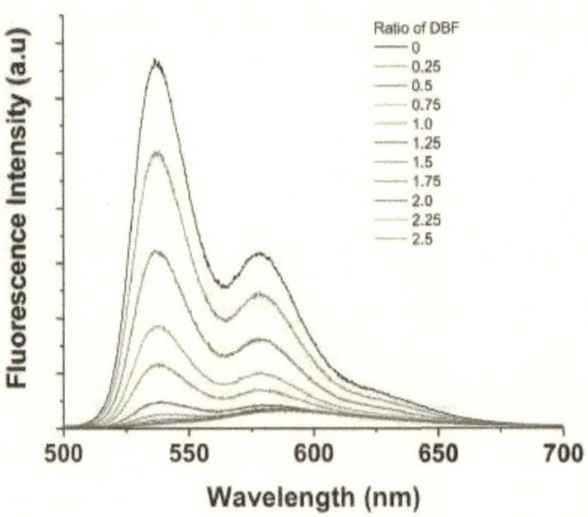

Figure 2.6: Decreasing fluorescence emission of PMI-CB[8] upon the addition of DBF at [c] = 4 μM in water (λ_{ex} = 480 nm).

The fluorophore and quencher guests were equipped with an amino terminated side chain as shown earlier in Figure 2.4. The two arms of the complex therefore react alongside the amino component of the epoxy resin during the curing process, with the epoxy functional groups of the bisphenol-A based component A. This results in the insertion of the complex into the matrix in the form of supramolecular cross-links. Upon curing, the two guests while now covalently linked to the resin as pendant groups, are held together by supramolecular interactions within the CB[8]. Amino terminated 4,7,10-trioxa-tridecane connectors were chosen for all guests with the dual purpose of increasing water solubility, necessary for ternary complex formation in water, and to impart conformational flexibility to permit their curing in the matrix. The loaded resin can be used to prepare composite materials reinforced by both glass and carbon fibre.

PER was synthesised based on a literature procedure starting from the perylene tetracarboxylic dianhydride (PTCDA), where the monopotassium salt of the dianhydride was reacted with a diamine glycol, as shown in Scheme 2.1.

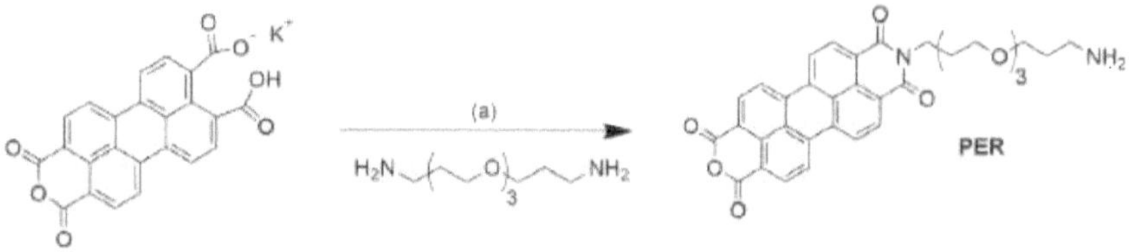

Scheme 2.1: Synthetic scheme for PER. (a) aq K_2CO_3,triethylamine, 90 °C, 48 h.

The synthesis of AZO was carried out through the corresponding bromo-azobenzene. Through an Ulmann coupling, the amino terminated chain was linked to the aromatic core.[27, 28]. DBF was synthesised by converting 3-dimethoxydibenzofuran amine to the corresponding isocyanate, followed by reaction with excess diaminoglycol.[29] The final synthetic steps are shown in Scheme 2.2. CB[8] was prepared and purified using a procedure adapted from literature, described in detail in Chapter 3.[30]

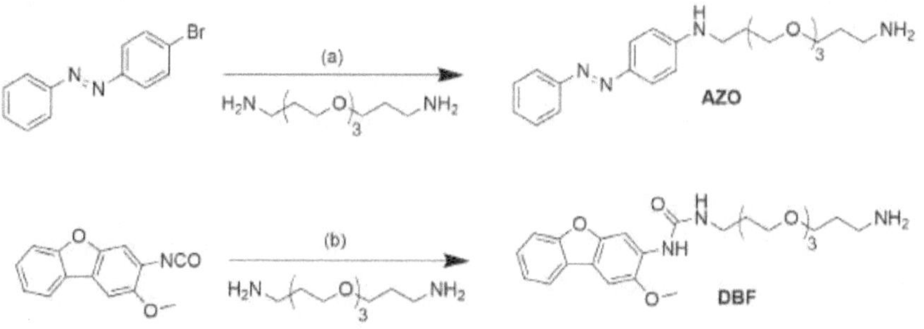

Scheme 2.2: Synthetic scheme for AZO and DBF. (a) L-proline, CuI, DMSO, 80 °C (microwave), 1 h; (b) DABCO (cat.), dry DCM, 12 h.

2.3.2   Preparation of self diagnostic composites

The strategy behind the design of the self-diagnostic composite requires the insertion of the reporting system as a complex in the polymer precursor before polymerisation in order to have fluorescence quenching of the reporting probe. This is carried out by linking the supramolecular complex to a commercial epoxy resin as shown in the schematic in Figure 2.7.

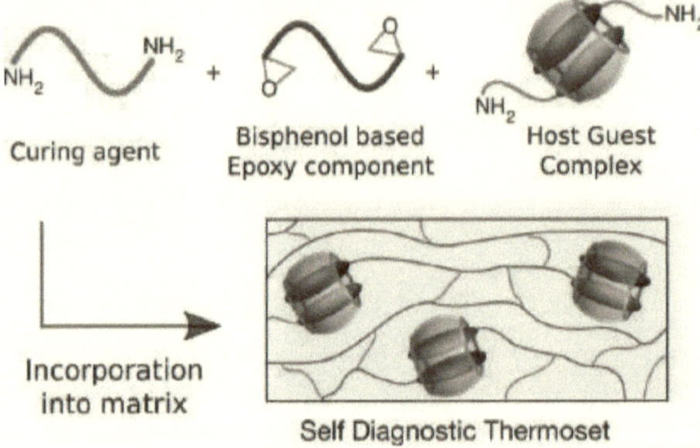

Figure 2.7: Components of the damage reporting carbon fibre-epoxy resin composite.

Commercial Elan-tech EC 157.1/W152 LR, a two-component epoxy system was used as the primary test matrix (provider Elantas Europe srl). It is a standard formulation for the preparation of carbon fibre composites via vacuum infusion and it has the advantage of room temperature curing with post curing at a moderate 50 °C - 60 °C range and is certified for use in high performance composite parts, commonly used for structural components such as boat hulls. The complexes are prepared in water using a 100-fold excess of quencher AZO and a 10-fold excess of DBF, to ensure the quantitative formation of the ternary complexes. The introduction of each ternary complex to the matrix was done in a step by step process as shown in Figure 2.8. An aqueous solution of the complex was first added to W152 LR, the amine component of the epoxy resin. In solution, a 100-fold excess of the second guest AZO was used to ensure the complete quenching of the PER, to give the complex perylene-azobenzene-CB[8] complex (PER-AZO-CB[8]). The concentration c[PER] in the matrix was $1 \times 10^{-6}\,\mathrm{mol\,kg^{-1}}$. The hardener W152 LR was first added to an aqueous solution of the complex to give a complex concentration of $4.3 \times 10^{-6}\,\mathrm{mol\,kg^{-1}}$, corresponding to 0.02 w% in the curing agent. The water was subsequently removed from the emulsion by heating at $100\,^{\circ}\mathrm{C}$ with constant agitation to ensure homogeneity. Upon removal of water the resulting curing agent W152 LR was cooled to room temperature, giving the loaded curing agent (LCA) containing the preformed CB[8] ternary complex.

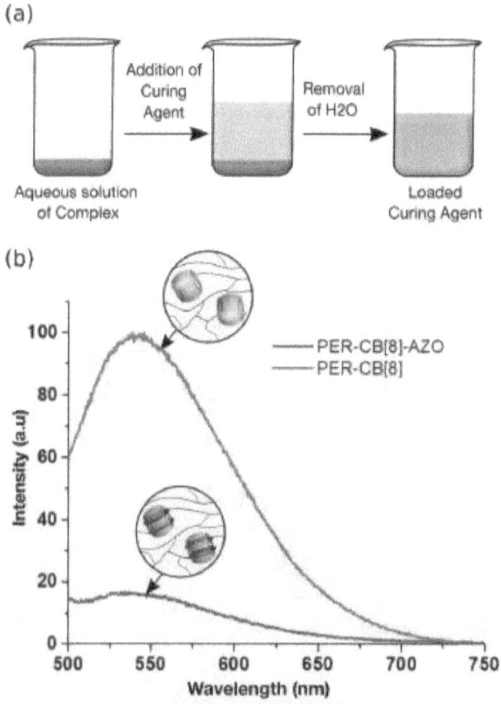

Figure 2.8: (a) Schematic representation of the incorporation of the complex into curing agent. (b) Emission spectra at $\lambda_{ex} = 480\,\mathrm{nm}$ of PMI-CB[8] (1:1) and PER-AZO-CB[8] (1:100:1) complexes in the polymer matrix

The stability and fluorescence emission of the system in the matrix was investigated by the preparation of samples of the thermoset containing the PER-AZO-CB[8] ternary complex and its PMI-CB[8] precursor in disposable polycarbonate cuvettes. The LCA was used in combination with the epoxy component EC 157.1 in the preparation of the thermoset using the curing procedure reported in the technical data sheet (Appendix-A). Figure 2.8(b) shows the fluorescence spectra of the cured epoxy resin with and without the AZO quencher. In the presence of the AZO quencher the emission of the PMI-CB[8] is drastically reduced with respect to the control sample containing PMI-CB[8] in 1:1 ratio. This also confirmed the stability of the complex to curing conditions of the epoxy resin. The same procedure was employed for the perylene-dibenzofuran-CB[8] complex (PER-DBF-CB[8]) and the control. The formation of the heteroternary complex is necessary for PER quenching in the matrix. The unlikely alternative of PER quenching by dispersed quencher in the matrix was ruled out by the following experiments: cured epoxy resins were prepared under standard conditions with LCA containing PER alone and PER plus an excess of DBF respectively. In both cases the fluorescence emission spectrum of PER remained unaltered (Figure A.2)

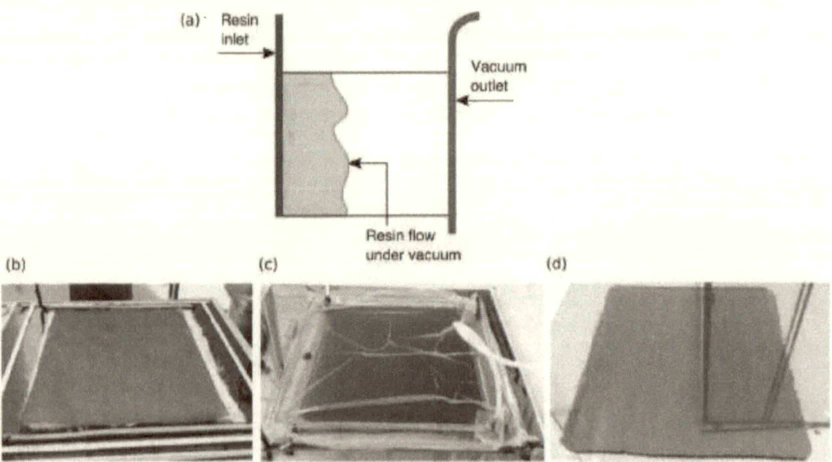

Figure 2.9: Schematic representation of (a) vacuum infusion process for preparation of the composite panel. (b), (c) pictures at various stages of infusion and (d) completed panel (500 mm × 500 mm × 3 mm)

The carbon fibre-epoxy panels were prepared via vacuum infusion of the loaded resin into carbon fibre fabric (Twill 2x2). The two components (EC157 and LCA containing PER-AZO-CB[8]) are mixed and the mixture is infused onto the carbon fibre fabric under vacuum (Figure 2.9b-d). The resulting panel was then cut to obtain specimens with direction of carbon fibre at 0° along the axis. The same procedure was followed in the case of PER-DBF-CB[8]. Blank panels were similarly prepared using the pure hardener containing no reporting agent. Standardised specimens were then prepared from the panel.

2.3.3 Tensile and compressive testing of self-diagnostic composite specimens

PER-AZO-CB[8]

The response of the PER-AZO-CB[8] containing composite to stress was studied by subjecting the specimens to uniaxial deformation testing in both tensile and compression modes in accordance with the American Society for Testing and Materials (ASTM) standards. They were subsequently examined under a fluorescence microscope to observe change in emission. In unstressed samples, images report a random position of the sample, while in stressed samples, images report a region of maximum fluorescence. For broken samples, images were acquired in correspondence to the breaking point. Both blank and unstressed specimen showed no fluorescence under the same conditions.

In detail, specimens containing the PER-AZO-CB[8] reporting agent were subject to tensile testing following the ASTM D3039 method. Three specimens were broken and two were subjected to stress equal to 60, 70 and 80% of the ultimate tensile strength (UTS) in order to check the response of the material prior to complete fracture as shown in figure 2.10(b).

The stressed specimen did not show any superficial damage compared to the pristine one. When viewed under a fluorescence microscope, a marked increase in background fluorescence was observed with a stress threshold above 70% as shown below in Figure 2.10(d). The fluorescence intensity increases with stress loading. No significant increase in fluorescence was detected below 70% (Figure S8). Upon breaking, fluorescence spots appear along the broken edge of the specimen (Figure 2.10(d)). The same behaviour was observed under uniaxial compression testing according to the ASTM D3410 method. In this case the final breaking can also follow a delamination pathway. Again, the fluorescence

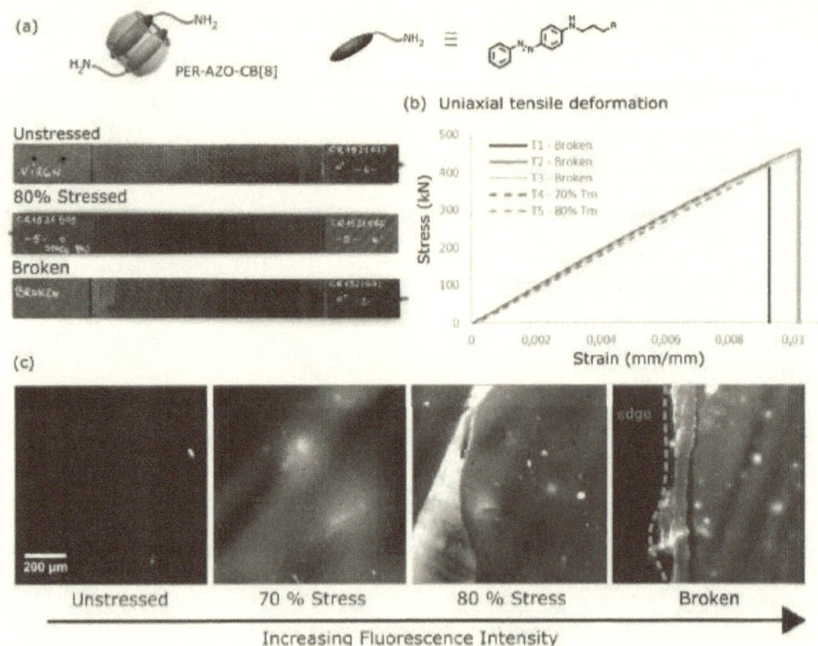

Figure 2.10: (a) Sketch of the employed PER-AZO-CB[8] reporting agent. (b) Image of Specimens under various conditions of stress (c) Stress strain curves for uniaxial tensile deformation (d) Fluorescence microscope images (10×) under varying degrees of tensile stress represented as percent of UTS.

appears at 70% stress, even if less strongly than in the tensile stress mode (Figure 2.11(c)).

In addition to tensile and compression, the specimens were also tested under flexion. In this case, the appearance of fluorescence in the damaged specimens was minimal. This can be attributed to the fact that this mode of mechanical testing causes least strain or deformation of the material. Additionally, in most experiments flexion led to a breaking of the specimen with a relatively low amount of force.

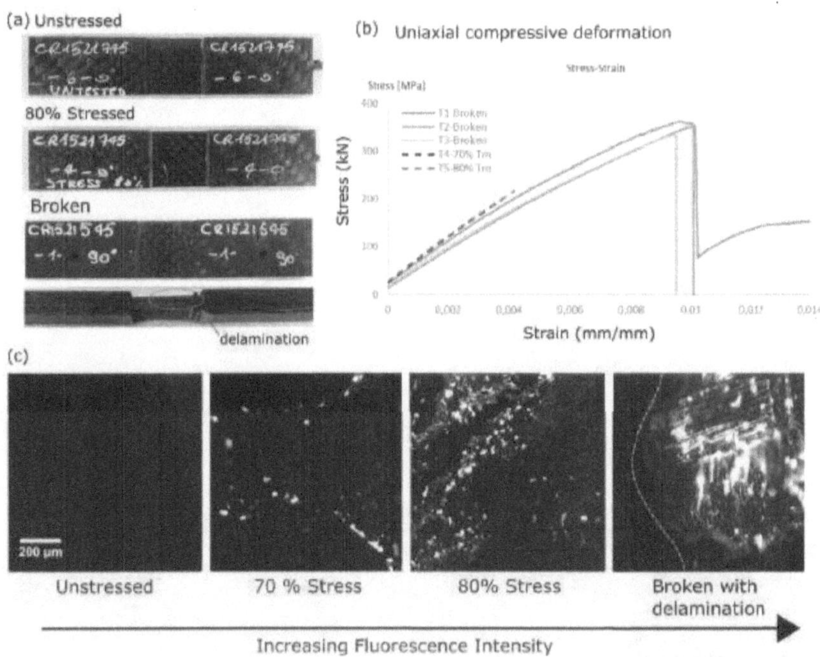

Figure 2.11: (a) Image of PER-AZO-CB[8] specimens under various conditions of stress (b) Stress strain curves for uniaxial compressive deformation (c) Fluorescence microscope images (10×) under varying degrees of tensile stress represented as percent of UTS.

PER-DBF-CB[8]

Exchanging the quencher in the ternary complex does not alter the self diagnostic properties of the resulting composite. The response of the self diagnostic composite comprising PER-DBF-CB[8] complex as the additive are comparable to those comprising PER-AZO-CB[8] both in terms of fluorescence intensity and stress sensitivity both in the tensile and compression modes (Figure 2.12). Though the diagnostic results are comparable, from a practical point of view, the PER-DBF-CB[8] complex is preferable to the PER-AZO-CB[8], since the system performs favourable with a lower quantity of the quencher guest. The lower the amount of additives, the less the interference with the mechanical properties of the matrix. This is important to consider for an additive based approach.

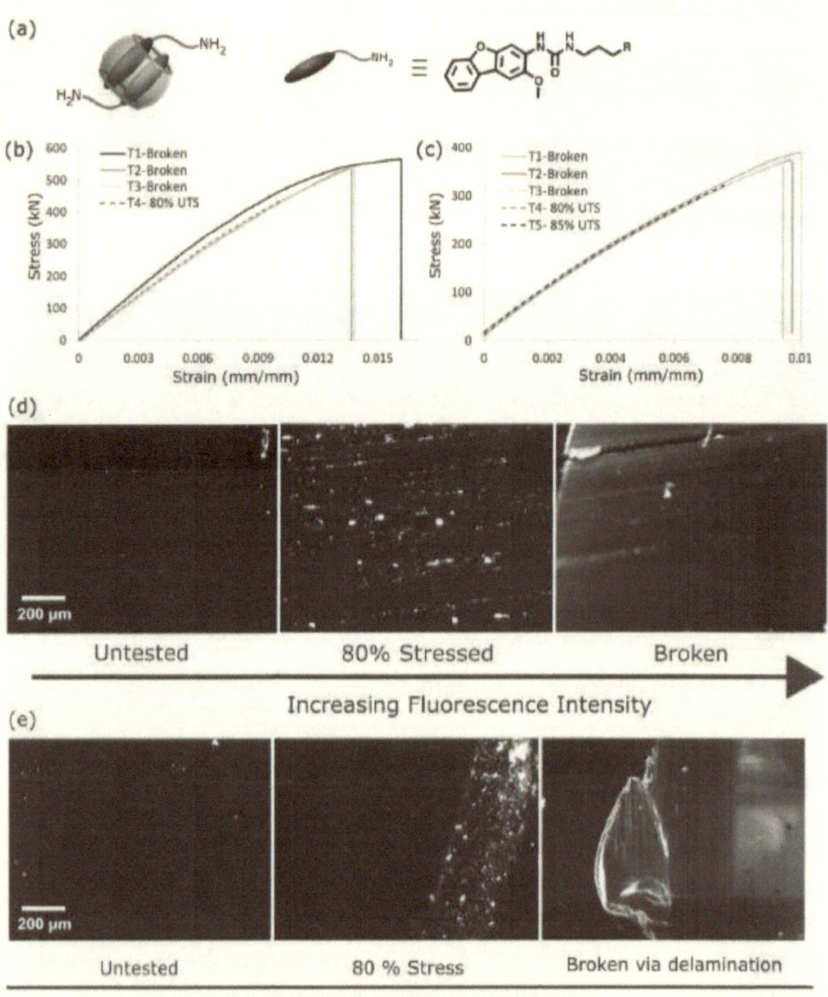

Figure 2.12: (a) Schematic representation of reporting complex PER-DBF-CB[8]. Stress strain curves of (b) uniaxial tensile deformation and (c) uniaxial compressive deformation of specimens containing the control probe. Corresponding fluorescence microscope images of specimens under varying degrees of (d) tensile and (e) compressive stress.

The optical resolution of strained regions is very high in both composites, with detectable fluorescent zones smaller than 20 µm. This is an important advantage of the use of turn-on fluorescence supramolecular probes, as it combines molecular level stress response with highly sensitive fluorescence strain detection. In addition, the fluorescence of the specimens was observed to persist over the course of months.

2.3.4   Control experiments

According to the proposed reporting mechanism, the crosslinking of the ternary complex within the matrix is pivotal to translate the localised strain into decomplexation-induced fluorescence. A control experiment was devised to test this tenet wherein the supramolecular complex would be distributed within the epoxy matrix of the carbon fibre epoxy composite but would not be covalently linked to the matrix as described in Figure 2.13.

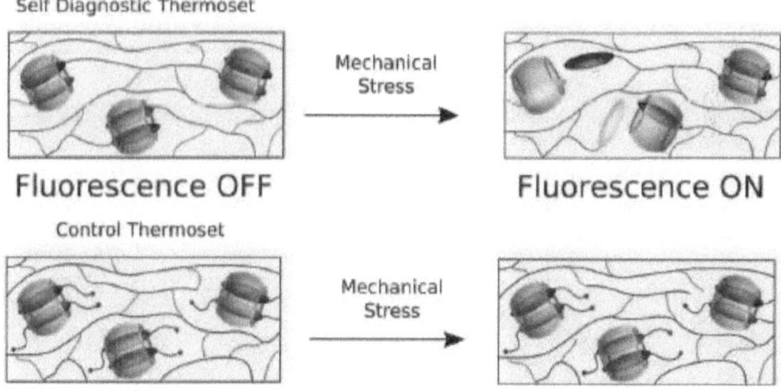

Figure 2.13: Schematic representation of control experiment

This was achieved by the synthesis of two molecules with identical structures to the guest molecules of the reporting complex except with both the reactive amino groups absent. The reactive groups of the PER fluorophore and the DBF quencher were substituted by methoxy units which remain inactive during the curing of the resin.

The two guests were prepared adapting known literature procedures.[26, 31] The behaviour of the control fluorophore PER-OMe in solution remains un-

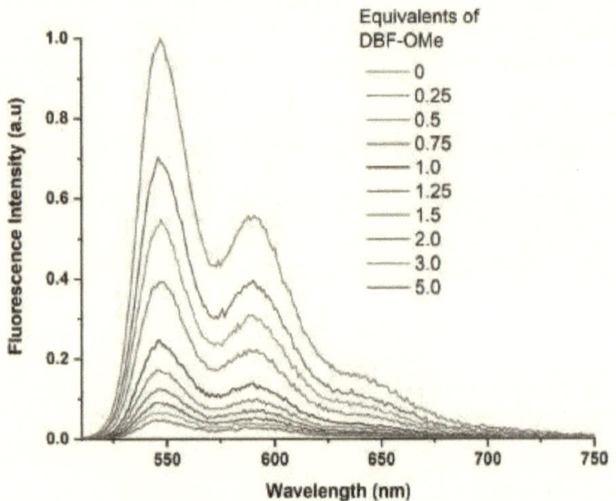

Figure 2.14: Decreasing fluorescence emission of PER-OMe-CB[8] upon the addition of DBF-OMe at [c] = 4 µM in water (λ_{ex} = 480 nm).

changed. An aqueous solution of PER-OMe-CB[8] in a ratio 1:1 was titrated with the control quencher DBF-OMe at a concentration of 4 µM. As expected the fluorescence emission decreases upon addition of the DBF-OMe as shown in Figure 2.14. Subsequently, this chemically inert ternary complex was incorporated into carbon fibre epoxy composites following the same procedure as for the self diagnostic system. The complex was randomly dispersed within the W152 LR curing agent to obtain the control LCA. The corresponding LCA was then used to prepare carbon fibre specimens via the same vacuum infusion protocol for mechanical testing. Uniaxial tensile and compressive testing experiments were carried out under the same conditions following which the specimens were observed under the fluorescence microscope. Very limited fluorescence spots appeared upon both uniaxial tensile and compression deformations compared to the corresponding cross-linked ternary complex PER-DBF-CB[8] (Figure 2.15).

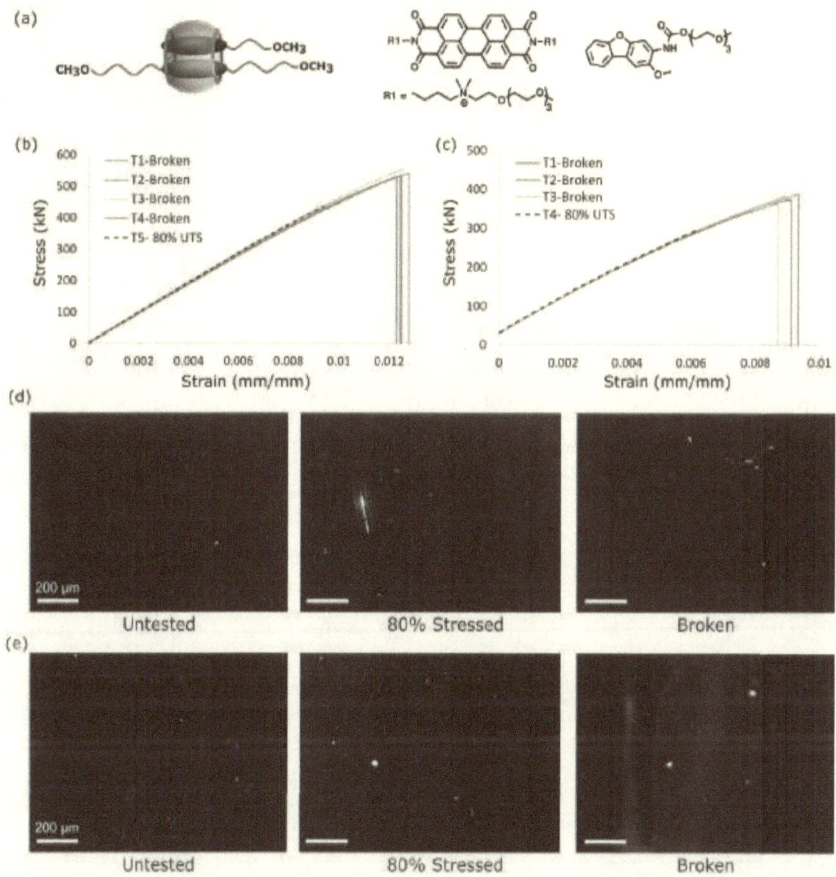

Figure 2.15: (a) Schematic representation of control complex. Stress strain curves of (b) uniaxial tensile deformation and (c) uniaxial compressive deformation of specimens containing the control probe. Corresponding fluorescence microscope images of specimens under varying degrees of (d) tensile and (e) compressive stress.

2.3.5 Response to fatigue damage

In addition to compressive and tensile testing, the specimens were subjected to fatigue testing to assess the performance of the material under conditions similar to actual use. Fatigue is the weakening of a material caused by repeated application of load.[30] If the loads are above a certain threshold, microscopic cracks begin to form. The stress values that cause such damage are significantly lower than the corresponding static ultimate tensile stress limit. Tensile fatigue testing was performed on composite specimens containing the **PER-DBF-CB[8]** reporting agent, cut to the size defined in ASTM D3479 (Figure 2.16(a)). The S-N (Stress-Number of cycles) or Wöhler diagram is a familiar way to represent the resistance of a given material to cyclic application of loads. It shows graphically the material stiffness reduction with repeated application of load in the form of the ratio of final elastic modulus/initial elastic modulus of the specimen (E_x/E_{x0}). Fatigue testing was carried out by progressively subjecting the specimens to 1000, 10,000 and 100,000 cycles under 60% UTS with a frequency of 10 Hz. The 60% UTS stress value was chosen because it is below the sensitivity threshold of the reporting agent (Figure A.3). Subjecting the specimen to 1000 cycles resulted in no stiffness drop, and no changes in fluorescence. After 10,000 cycles few fluorescence spots appeared in concordance with an extremely limited stiffness drop (Figure 2.18, experimental section) Finally, after 100,000 cycles, a 40% stiffness drop was observed and fluorescence was visible along the entire length of the specimen (Figure 2.16(c)). It is particularly relevant to note the significant increase in fluorescence that appear at 10,000 cycles. At this stage, the stiffness drop is low and the specimen is visually intact. The appearance of fluorescence in this specimen confirms the ability the system to identify damage prone areas prior to failure of the material.

Carbon fibre laminate fabrics are made of many threads woven on a warp and a weft at 0° and 90° yarns orientation.[32] The loading is usually applied in one of the two fibre directions. In our case, monotonically increased tensile load was applied along the 0-direction, resulting in multiple ply microcracks in the 90-plies (Figure 2.16(g-h). Cross section micrographs of the specimens post fatigue testing are shown in Figure 2.17. The image shows carbon fibres oriented in the 90 and 0 degree direction and microcracks caused by fatigue are clearly visible. Fluorescence in the 0° orientation was observed only after significant stiffness drop (Figure 2.16(i)), as an indication of extensive damage. This is significantly important since the presence of microscopic damage in the 0°

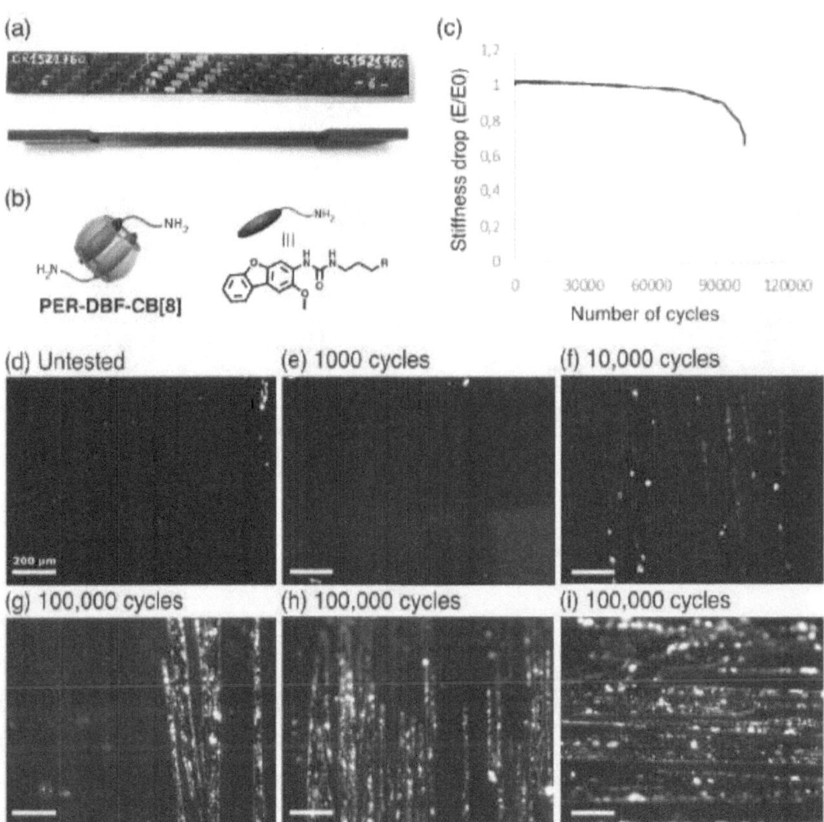

Figure 2.16: Fatigue testing with PER-DBF-CB[8] reporting agent: (a) Specimen image after complete fatigue testing showing the visual integrity of the specimen (b) Structure of the ternary complex in the specimen (c) S-N diagram showing normalised stiffness drop (E_x/E_{x0}) with increasing number of cycles (d-i) fluorescence microscope images (10×) under increasing cycles of stress of 60% UTS: (d) pristine specimen; (e) after 1000 cycles; (f) 90° fluorescence emission after 10,000 cycles; (g) and (h) 90 deg fluorescence emission after 100,000 cycles; (i) 0 deg fluorescence emission after 100,000 cycles

orientation is the harbinger of catastrophic failure in a composite. The spatial resolution attainable for detecting microcracks is in the low micrometer range, well below the resolution achievable with highly sensitive NDT techniques such as Phased Array Ultrasonic Testing (PAUT).[33]

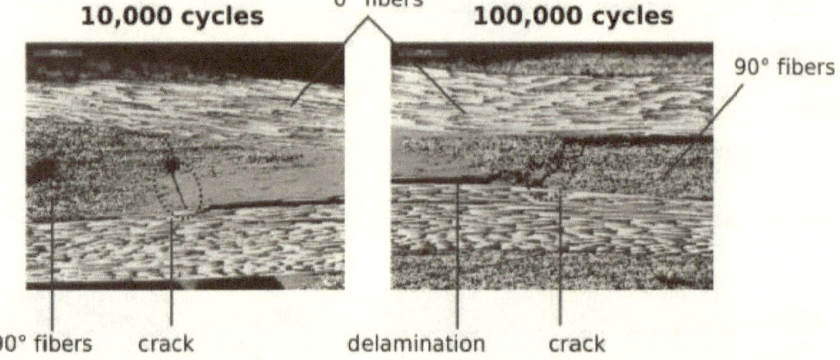

Figure 2.17: Cross Section Micrographs of specimens subjected to 10,000 and 100,000 cycles of 60% UTS (10 Hz).

2.4 Conclusions

In summary, we report a novel system for early stage damage detection in a carbon fibre reinforced composite by incorporating supramolecular crosslinks within the matrix facilitated by the host molecule CB[8]. CB[8] encapsulates two molecules in the matrix, a fluorophore and a quencher forming a ternary complex wherein the emission of the probe is suppressed. The epoxy matrix was used to prepare carbon fibre epoxy composite panels as per standard industrial methods, which were subjected to mechanical deformation testing. Under the application of stress, the weak supramolecular links within the matrix break apart, and the fluorescence of the probe is reinstated, assisting in the visualisation of microscopic damage in the first layer of the composite material. Moreover, samples subjected to increasing stress show a commensurate increase in fluorescence intensity. Correlation of the compressive and tensile tests with emergence of fluorescence at 70% of UTS demonstrates that the proposed reporting agents are capable of detecting early damage in carbon fibre epoxy composites. In addition to tensile and compressive stress, the system shows the ability to detect fatigue damage.

Exploiting the dissociation of noncovalent interactions in combination with turn on fluorescence has enabled self-diagnosis in composite materials at 0.003 w% of the reporting systems. This additive-based approach allows to directly incorporate the damage-reporting probe into a commercial polymer without altering its mechanical properties (Figure A.5). In addition, the persistence of the fluorescent signal with time allows the detection of affected regions long after the first appearance of damage. Overall, this approach offers an innovative, technologically viable solution to the quest of simple and effective methods of monitoring the structural integrity in composites. The proposed approach can be applied to a broad range of thermosets and related composites.

2.5   Experimental section

Materials and methods

Materials. All starting materials were purchased from Sigma Aldrich, TCI chemicals, Alfa Aesar and Fluorochem and used as received.

Nuclear magnetic resonance (NMR). ^{1}H$-$NMR spectra were recorded on a JEOL-ECZ600R (600 MHz), Bruker Avance-400 (400 MHz), spectrometer or a Bruker Avance-300 (300 MHz) at room temperature as specified. Chemical shifts (δ) are reported in parts per million (ppm). ^{1}H$-$NMR chemical shifts are given in reference to the residual solvent peaks. ^{13}C$-$NMR spectra were obtained using a Bruker AVANCE 300 (75 MHz) or a Bruker AVANCE 400 (100 MHz) spectrometer. All chemical shifts (δ) were reported in ppm relative to the carbon resonances of the NMR solvents.

Fluorescence measurements. Samples for fluorescence characterisation were prepared by diluting a stock solution to get the desired sample concentrations. Spectra were recorded on a fluorescence spectrophotometer (Horiba JobinYvon FluoroMax-3) at room temperature (25 °C).

Electrospray Ionization Mass Spectrometry (ESI-MS). MS Measurements Electrospray ionization ESI-MS experiments were performed on a Waters ZMD spectrometer equipped with an electrospray interface.

Mechanical testing. Samples were tested in accordance with American Society for Testing and Materials (ASTM) standards.

Tensile Testing under ASTM D3039. Specimens were subjected to uniaxial tensile testing using MTS Insight Electromechanical Testing Systems (Figure in appendix A) 150 kN, with 250 kN hydraulic grips at 2 mm/min (Grip pressure: 100 bar) following the ASTM D3039 method. This test method is designed to produce tensile property data for material specifications, research and development, quality assurance, and structural design and analysis. The method determines the in-plane tensile properties of polymer matrix composite materials reinforced by high-modulus and high-strength fibres.

Compression Testing under ASTM D3410 Specimens were subjected to compression testing using MTS Insight Electromechanical Testing Systems (Figure in appendix A) 150 kN, with 250 kN hydraulic grips at 2mm/min (Grip pressure: 100 bar) following the ASTM D3410 method. This test method determines the in-plane compressive properties of polymer matrix composite materials reinforced by high-modulus and high-strength fibres.

Fatigue Testing Tensile fatigue specimens were cut to size according to ASTM D3479 (length 150 mm; width 25 mm). The mechanical tests were performed using MTS Landmark 100 kN.

Fluorescence imaging measurements. Fluorescence measurements were performed with a Nikon Eclipse Ti inverted microscope, equipped with an Intensylight Epifluorescence Mercury illuminator and a FITC fluorescence cube ($\lambda_{ex} = 480 \pm 30$ nm , $\lambda_{em} = 535 \pm 45$ nm). Images were acquired by an Andor Clara high sensitivity 16-bit camera. A $10\times$ objective was used; all fluorescence images reported in the paper have size $936 \times 936\,\mu m^2$. Exposure time was set at 0.5 seconds; to allow image comparison, their intensity scale was fixed within a 700-7000 count range, chosen to improve readability. Image processing was done using Micro-Manager 1.4.22.

Synthesis

Perylene-3,4,9,10-tetracarboxylic dianhydride monopotassium salt (1)

1 was prepared by treating a solution of PTCDA (10 g, 25.5 mmol) in water (800 mL) with KOH (40 g, 0.71 mmol) and stirring the resulting mixture at 90 °C for 2 h. Acetic acid (50 mL) was then added and the reaction was stirred at 90 °C for 40 min. The precipitate (10 g, 93%) was removed by filtration, washed with methanol and dried at 120 °C. The compound was used without further purification.

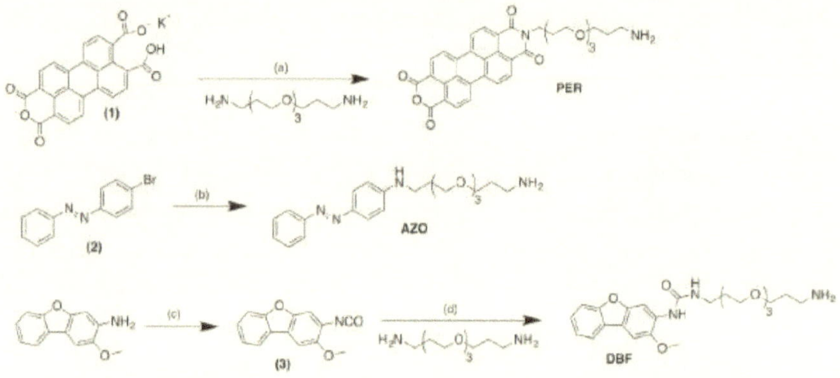

Scheme 2.3: Synthetic scheme for guest molecules: (a) aq K_2CO_3, triethylamine, 90 °C, 48 h, 65%; (b) (c) 4,7,10- trioxa-1,13-tridecanediamine, L-proline, CuI, DMSO, 80 °C (microwave), 1 h, 68% and DBF (d) triphosgene, triethylamine, dry DCM, r.t., 12 h, 75% (e) DABCO (cat.), dry DCM, 12 h, 90%

Perylene monoimide (PER)

3,4,9,10-Perylenetetracarboxy-3,4-anhydride monopotassium salt **1** (0.96 g, 2 mmol) and 4,7,10-trioxa-1,13-tridecanediamine (2.37 g, 10 mmol) were placed in a round bottom flask and water (40 mL) was added. The solution was stirred at 90 °C for 48 h before the addition of aqueous potassium carbonate (25 % w/w, 100 mL). The solution was heated at 90 °C for 3 h, over which time the color changed from purple to green. The solid was filtered off and washed from the filter with a water (150 mL) and triethylamine (5 mL) mixture. The filtrate was diluted with HCl (2 M, 250 mL) and after sitting overnight the precipitated solid was filtered off and washed with methanol. The product was obtained as a purple solid. (0.77 g, 65 %)

^{1}H−NMR (400 MHz, DMSO-d$_6$) δ: 8.80 (d, J = 7.9 Hz, 2H), 8.47 (dd, J = 11.7, 8.0 Hz, 2H), 7.75-7.69 (m, 4H), 4.16-4.13 (m, 2H), 3.55-3.43 (m, 4H), 3.09 (dd, J = 7.1, 4.9 Hz, 2H), 2.83 (td, J =6.4, 0.6 Hz, 2H), 1.93 (dd, J = 7.1, 6.4 Hz, 2H), 1.77 (dd, J = 13.2, 6.4 Hz, 2H);

MS (ESI) m/z: [M + H]+ calcd for $C_{34}H_{31}N_2O_8$, 595.21; found, 595.36.

(E)-1-(4-bromophenyl)-2-phenyldiazene (2)

2 was prepared according to a literature procedure.[27] A solution of H_2O_2 (35%, 3.3 mL) in 2.7 mL water was added to a suspension of 4-bromoaniline (1 g, 5.81 mmol) in MeOH (1.8 mL). MoO_3 (90 mg) and 1 M KOH (0.6 mL) were added. The reaction mixture was stirred for 48 h at room temperature, the precipitate was obtained by filtration washed with water and dried to give the 1-bromo-4 nitrosobenzene without further purification. Aniline (41.2 mg, 4.5 mmol) was added to a solution of the above 1-bromo-4-nitrosobenzene in acetic acid (60 mL) and the mixture was refluxed for 12 h. The solvent was evaporated and the crude product was purified via column chromatography on silica gel (Hexane:DCM 95:5) to give 2 in 77% yield.

^{1}H−NMR (300 MHz, $CDCl_3$): δ 7.95-7.92 (dd, 2H), 7.85-7.81 (dd, 2H), 7.70-7.65 (dd, 2H), 7.58-7,53 (dd, 2H), 7.51-7.49 (t, 1H).

ESI-MS $[M+H]^+$ m/z calc for $C_{12}H_9BrN_2$: 261.12 ; found, 261.00.

AZO

A solution of (E)-1-(4-bromophenyl)-2-phenyldiazene **2** (52 mg, 0.2 mmol), L-proline (46 mg, 0.4 mmol), copper(I) iodide (37 mg, 0.2 mmol), Na2CO3 (31 mg, 0.3 mmol) and 4,7,10-trioxa-1,13-tridecanediamine (1.75 mL, 7.9 mmol) in DMSO (2.3 mL) was heated under microwave irradiation for 1 h at 80 °C. DMSO and 4,7,10-trioxa-1,13-tridecanediamine were removed from the reaction mixture and the remaining crude purified by chromatography (DCM:MeOH : 9:1), yielding an orange solid (54 mg, 68%).

^{1}H−NMR (300 MHz, CD_2Cl_2) δ: 7.86-7.82 (m, 4H), 7.54-7.49 (m, 2H), 7.45-7.36 (m, 1H), 6.74−6.71 (m, 2H), 3.69−3.58 (m, 14H), 3.37 (t, J=6.0 Hz, 2H), 2.88 (bs, 1H), 2.22 (bs, 2H), 2.01−1.93 (m, 2H), 1.77 (t, J=6.0 Hz, 2H), 1.16-1.14 (m, 2H);

ESI-MS m/z: $[M + H]^+$ calc for $C_{22}H_{33}N_4O_3$, 401.25; found, 401.26.

3-isocyanato-2-methoxydibenzo[b,d]furan (3)

A solution of 2-methoxydibenzo[b,d]furan-3-amine (0.5 g, 2.3 mmol) and triethylamine (0.7 mL, 5.2 mmol) in 50 mL anhydrous DCM was cooled to 0 °C under inert atmosphere and treated with triphosgene (1.5 g, 5.2 mmol). The mixture was stirred overnight at room temperature and purified by chromatography over silica gel. The solvent was removed to give the product (0.4 g, 75%) as a solid.

^{1}H−NMR (400 MHz , CDCl$_3$) δ: 7.90 (d, 1H), 7.57 (d, 1H), 7.48 (m, 1H), 7.38 (m, 2H), 7.25 (s, 1H), 4.03 (s, 3H).

DBF

To a solution of 4,7,10-trioxa-1,13-tridecanediamine (1.2 mmol) in dry DCM, a solution of **3** (0.14 g, 0.6 mmol) was added drop-wise and the mixture was stirred for 12 h at room temperature in the presence of a catalytic amount of DABCO. The solvent was removed, and the product was purified via preparative TLC to give DBF (90%).

^{1}H−NMR (400 MHz, CDCl3): δ 8.49 (s, 1H), 7.78-7.76 (bs, 1H), 7.48 (m, 2H), 7.34 (bs, 1H), 7.28-7.24 (m, 2H), 5.98 (bs, 1H), 3.92 (bs, 3H), 3.71-3.61(m, 12H), 3.45-3.42 (m, 4H), 1.87-1.81 (m, 4H), 1.50 (bs, 2H).

ESI-MS m/z: $[M + H]^+$ calc for $C_{24}H_{34}N_3O_6$, 460.24; found, 460.25.

Perylenediimide (4)

4 was prepared adapting a literature procedure.[31] A suspension of PTCDA (1 g, 2.5 mmol) and 2-dimethylaminopropylamine (3 mL, 23.84 mmol) in 10 mL DMF was stirred for 6 h at 130 °C. The reaction mixture was cooled to room temperature and the product was precipitated by the addition of excess THF. The precipitate was washed with THF and dried in vacuum to give **4**. (1.28 g, 90%).

^{1}H−NMR (300 MHz, DMSO-d$_6$): δ 8,63 (bd, 4H), 8,38 (bd, 4H), 4,15 (t, 4H), 3,23 (t, 4H), 2,81 (s, 12H), 2,11 (m, 4H).

PER-OMe

A suspension of **4** (See Scheme 2.4) (0.1 g, 0.178 mmol) in 1-bromo-4-methoxy-triethyleneglycol (0.5 g, 1.78 mmol) was stirred at 100 °C overnight. The reaction was cooled to room temperature and 30 mL of THF were added under

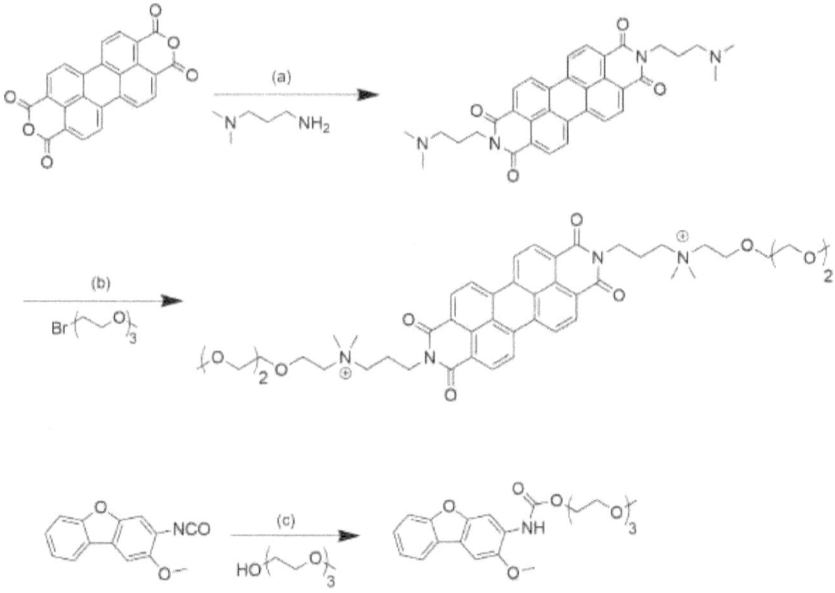

Scheme 2.4: Synthesis of control guest molecules PER-OMe (a) 130°C, DMF, 5 h, 85%; (b) 100°C, 12 h, 92% DBF-OMe (c) DBTDL (cat), r.t., 12 h, 83%.

stirring. The precipitate formed was filtered and washed with THF. The precipitate was dissolved in water and filtered. Solvent was removed via rotary evaporation to afford PER-OMe as a purple sticky solid (0.165 g, 92%).

^{1}H$-$NMR (300 MHz, DMSO-d$_6$): δ 8.85 (bs, 4H); 8.51 (bs, 4H), 4.15 (t, J=7.4 Hz, 4H), 3.82 (t, J=7.4 Hz, 4H), 3.52-3.46 (m, 24H), 3.20 (s, 6H), 3.04 (s, 12H), 2.17(m, 8H);

ESI.MS m/z [M]$^{2+}$ calc for $C_{48}H_{62}N_4O_{10}$, 427,52, found 427.41.

DBF-OMe

To a solution of 3-isocyanato-2-methoxydibenzo[b,d]furan (3) in 5 mL anhydrous dichloromethane, 4,7,10-trioxa-1,13-tridecanediamine (0.07 mL, 0.42 mmol) and a catalytic amount of DBTDL was added. The solution was stirred at ambient temperature for 12 h. The solvent was removed by rotary evaporation and the product was purified by elution over silica gel (DCM:MeOH 9:1) to afford the product (0.14 g, 83%).

^{1}H−NMR (300 MHz, CDCl$_3$): δ 8.40 (s, 1H), 7.83 (d, 1H, J = 6 Hz), 7.59 (bs, 1H), 7.53 (d, 1H, J = 6 Hz), 7.40-7.35 (m, 1H), 7.35 (s, 1H), 7.30-7.26 (m, 1H), 4.38 (t, J = 3 Hz, 2H), 3.98 (s, 3H), 3.78 (t, J = 3 Hz, 2H), 3.71-3.55 (m; 6H), 3.57-3.55 (m, 2H), 3.37 (s, 3H).

ESI-MS m/z [M + H]+ calc for C$_{21}$H$_{26}$NO$_7$, 404.17; found 404.36.

Preparation of LCA

LCA containing PER-AZO-CB[8]

Stock solutions of PMI-CB[8] at [c] = 65 µM and AZO at [c] = 5 mM were prepared in millipore water. To 20 mL of PMI-CB[8] stock solution, 26 mL of AZO stock solution was added, with the ratio of components in solution CB[8]:PER:AZO = 1:1:100. This solution is added to 300 g of the hardener W152LR. The obtained emulsion was placed in the oven at 100 °C and homogenised until the water was completely removed to give LCA loaded with PER-AZO-CB[8]. The loaded sample was removed from the oven and cooled to room temperature. The component containing the PER-AZO-CB[8] is subsequently used for curing the epoxy resin.

LCA containing PER-DBF-CB[8]

20 mL of PER CB[8] stock solution, was used to dissolve 6 mg of DBF with the ratio of components in solution CB[8]:PER:DBF = 1:1:10. The solution was used to prepare the loaded curing agent using the above procedure.

LCA containing PER

A solution of PER at [c] = 65 µM was prepared in milliQ water. 0.2 mL of PER stock solution was added to 3 g of the hardener W152LR. The obtained emulsion was placed in the oven at 100°C and homogenised until the water was completely removed to give LCA loaded with PER. The loaded sample was removed from the oven and cooled to room temperature. The component containing the PER is subsequently used for curing the EC 157.1 epoxy resin (10 g).

LCA containing PER and DBF

Solutions of PER at [c] = 65 µM and DBF at [c] = 65 µM were prepared in millipore water. 0.2 mL of PER stock solution and 2 mL of DBF stock solution were added to 3 g of the hardener W152LR. The obtained emulsion was placed in the oven at 100 °C and homogenized until the water was completely removed to give LCA loaded with PER + DBF. The loaded sample was removed from the oven and cooled to room temperature. The component containing the PER + DBF is subsequently used for curing the EC 157.1 epoxy resin (10 g).

Preparation of self diagnostic carbon specimens

For the preparation of composite specimens, a commercial epoxy system by Elantas Europe Srl was employed (EC157.1 epoxy resin and W152LR hardener). The mixing ratio for EC157.1-W152LR is 100:30 by weight. The appropriate quantity of hardener W152LR containing the ternary complex incorporated via the procedure described above was added to EC157.1 and mixed carefully. Additional information regarding resin curing is provided in the technical data sheet. Carbon fibre epoxy composite panels were prepared by vacuum infusion into twill carbon fibre fabric as shown in Figure ??

In order to fabricate a Panel of 500x500 mm with a thickness of 3 mm the following ratio of components was used.

- 833 g of component A EC157.1

- 250 g of component B W152LR (Curing agent)

- 12 sheets of Twill Carbon fibre fabric weighting 600 g

Fatigue Testing under ASTM D3479

Tensile fatigue specimens were cut to size according to ASTM D3479 (length 150 mm; width 25 mm). The mechanical tests were performed using MTS Landmark 100 kN. The untested specimen showed no significant fluorescence when viewed under a fluorescence microscope. The specimen was consecutively subjected to 1000 and 10,000 cycles of 60% of UTS with a frequency of 10 Hz.

A low stiffness drop was observed. The specimen was then subject to 100,000 cycles under the same conditions. A 40% of stiffness drop was observed after 100,000 cycles.

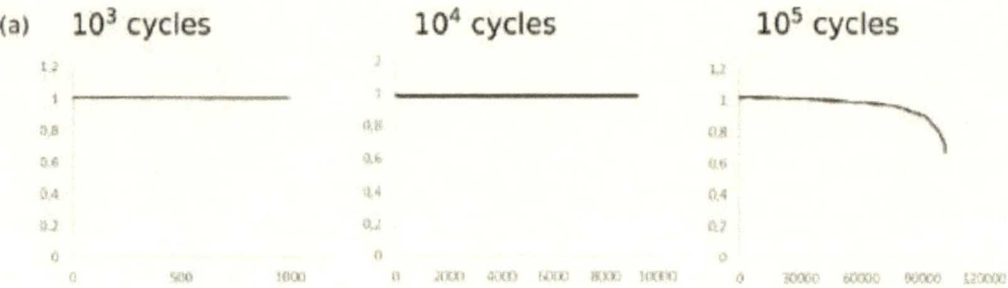

Figure 2.18: Wohler curves for 1000, 10,000 and 100,000 cycles of stress.

Appendix A

A.1 ITC binding curve for formation of ternary complex

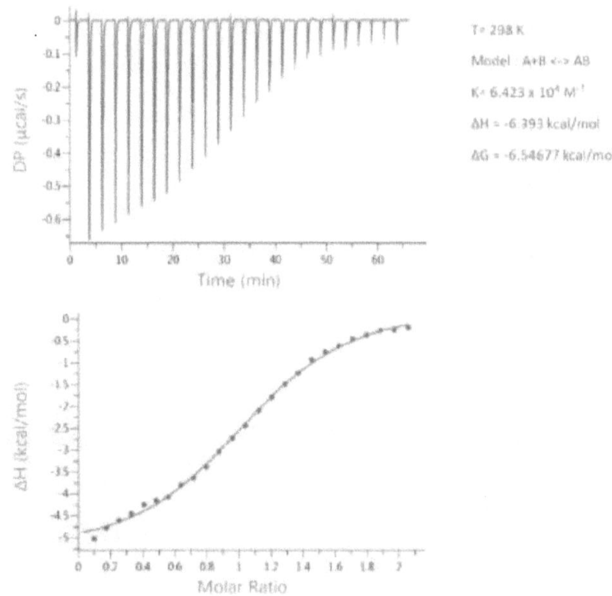

Figure A.1: ITC data for the titration of PER-CB[8] (0.1 mM) with AZO (1 mM) in water at 25 °C

A.2 Fluorescence spectra in the absence of CB[8]

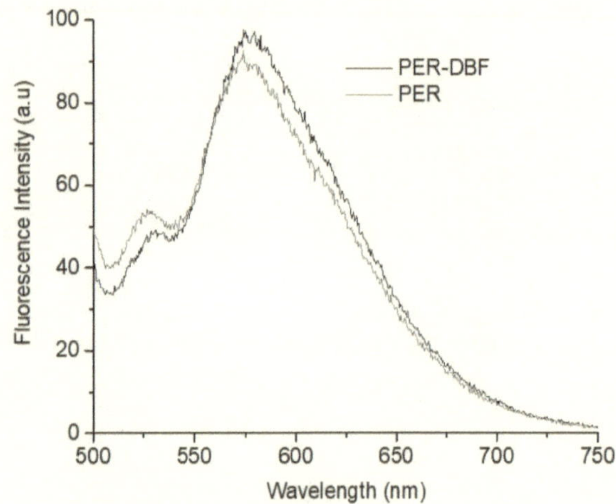

Figure A.2: Fluorescence emission spectra of cured epoxy resins EC157.1 with LCA containing PER (red trace) and PER + DBF 1:10 (black trace) (λ_{ex}= 480 nm) in polycarbonate cuvettes.

A.3 Fluorescence microscope images below 60% UTS

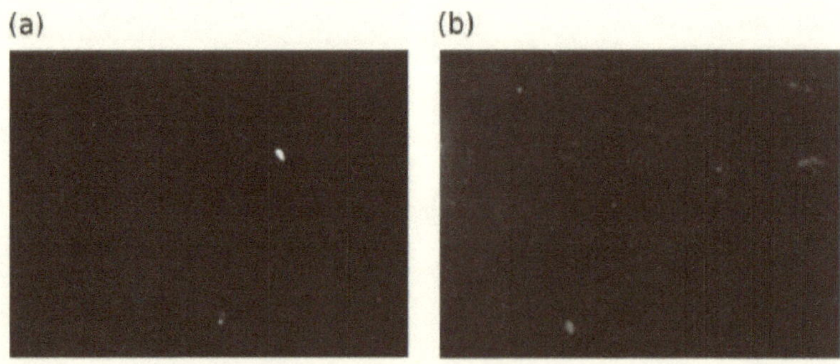

Figure A.3: (a) Untested specimen containing PER-AZO-CB[8]; (b) the same specimen after uniaxial tensile stress at 60% UTS. Photographs show a 936 × 1036 µm^2 section of the specimen

A.4 Mechanical testing

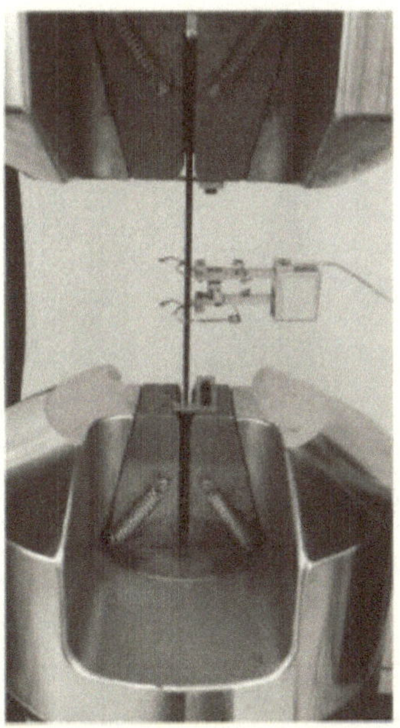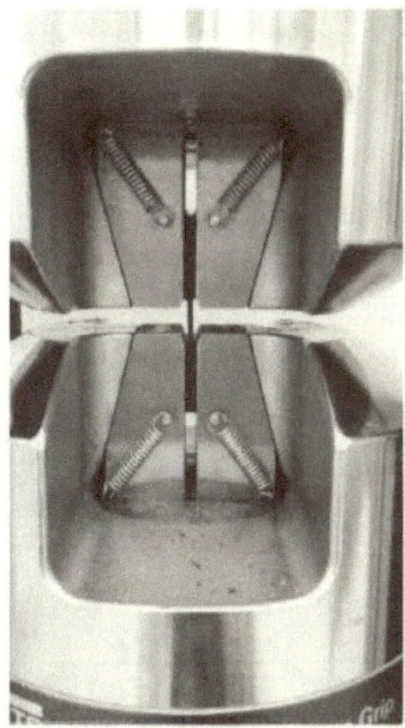

Figure A.4: MTS Insight electromechanical testing system used for uniaxial tensile (left) and compressive deformation testing (right).

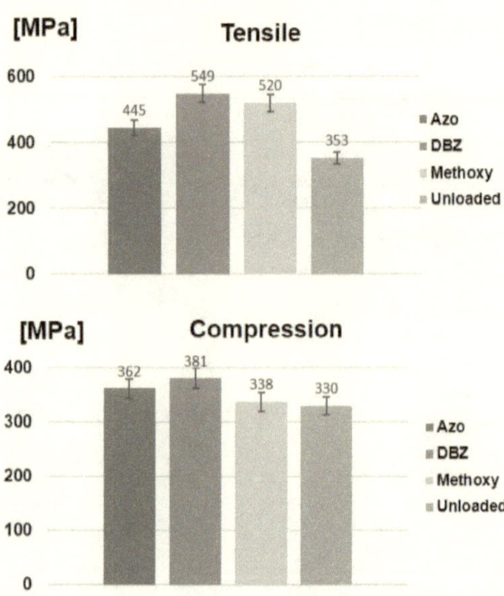

Figure A.5: Comparison of mechanical properties of the four different composite materials tested via a) uniaxial tensile and (b) compressive deformation testing. Azo: PER-AZO-CB[8] composite; DBF: PER-DBF-CB[8] composite; Methoxy: control reporting complex composite; Unloaded: standard composite.

A.5 Testing of self diagnostic system in glass fibre reinforced composites

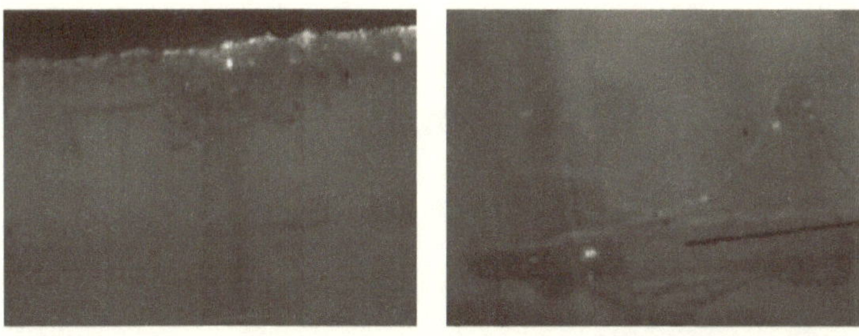

Figure A.6: Untested: Fluorescence only around cut edge

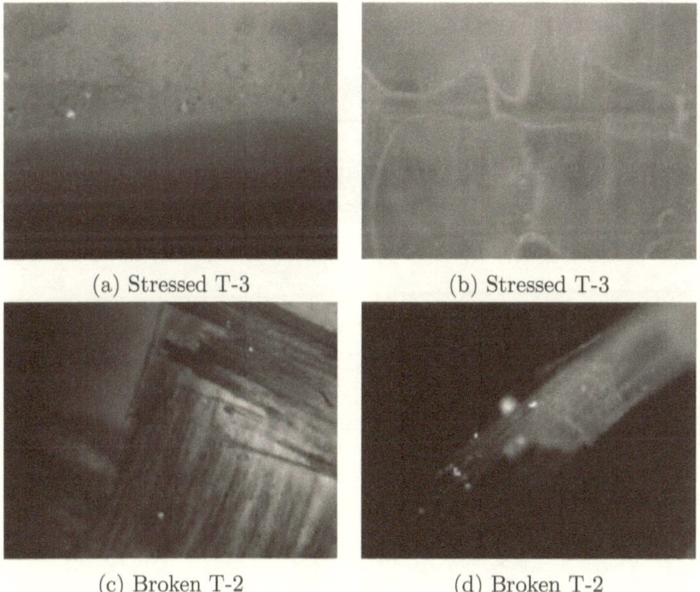

(a) Stressed T-3

(b) Stressed T-3

(c) Broken T-2

(d) Broken T-2

Figure A.7: Stressed (T-3) and broken (T-2) specimens under fluorescence microscope. Photographs show a $936 \times 1036\,\mu m^2$ section of the specimen.

A.6   Data sheet for EC 157.1-W152LR

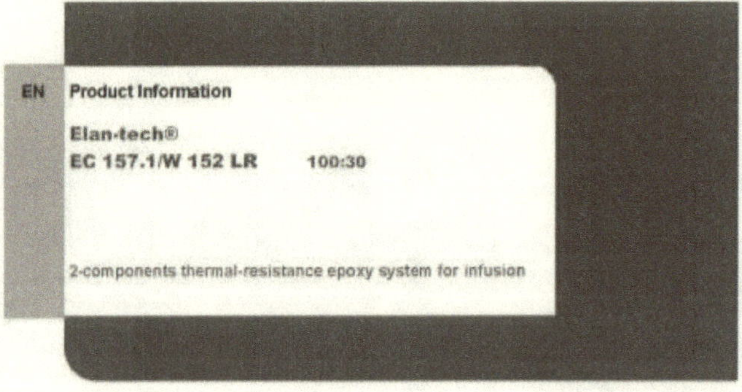

Resin	Hardener	Mixing ratio by weight
EC 157.1	**W 152 LR**	**100:30**

Application: High performance composite parts of medium size.

Processing: Manual mixing. Mechanical mixing. Mechanical mixing with automatic mixing/dispensing machines. Impregnation by infusion or under vacuum infusion (SCRIMP) of glass, carbon, kevlar fabrics. Room temperature curing.
W 152.1 HR: High reactivity for small components or as accelerator for other hardeners (see Technical Data Sheet of EC 157.1/W 152.1 HR).
W 152 MR: Medium reactivity for medium-small size components (see Technical Data Sheet of EC 157.1/W 152 MR).
W 152 LR: Medium-slow reactivity for medium-large size components.
W 152 MLR: Medium-slow. Medium and large size components (see Technical Data Sheet of EC 157.1/W 152 MLR).
W 152 XLR: Long pot life. Large size components (see Technical Data Sheet of EC 157.1/W 152 XLR).

Description: Two components epoxy system. Low viscosity. Good thermal resistance. Curing at room temperature plus the post-curing at moderate temperature (50-60°C) allows to obtain high performances.
The system is RoHS compliant (European directive 2002/95/EC) and the new RoHS Directive 2011/65/EU (RoHS 2) entered into force on 21 July 2011 and requires Member States to transpose the provisions into their respective national laws by 2 January 2013.

SYSTEM SPECIFICATIONS

Resin

Viscosity at:	25°C	IO-10-50 (ISO3219)	mPas	500	700

Hardener

Viscosity at:	25°C	IO-10-50 (ISO3219)	mPas	20	40

TYPICAL SYSTEM CHARACTERISTICS

Processing Data

				Colourless	
				Pale/yellow	
Colour resin				Colourless	
Colour hardener				Pale/yellow	
Mixing ratio by weight		for 100 g resin	g	100:30	
Mixing ratio by volume		for 100 ml resin	ml	100:37	
Density	25°C Resin	IO-10-61 (ASTM D 1475)	g/ml	1,13	1,17
Density	25°C Hardener	IO-10-61 (ASTM D 1475)	g/ml	0,93	0,97
Pot life	25°C (50mm;200ml)	IO-10-63 (*)	min	100	120
Exothermic peak	25°C (50mm;200ml)	IO-10-63 (*)	°C	180	195
Initial mixture viscosity at:	25°C	IO-10-50 (ISO3219)	mPas	150	250
Gelation time	25°C (1mm)	IO-10-69 (ASTM D8995-01)	h	10	12
Demoulding time	25°C (15ml;6mm)	(*)	h	24	32

PRODUCT INFORMATION   pag.2/4   ⬤ **ELANTAS**

EC 157.1/W 152 LR

TYPICAL CURED SYSTEM PROPERTIES

Properties determined on specimens cured: 24 h RT + 15 h 60°C

Colour				Pale yellow	
Machinability				Excellent	
Density 25°C		IO-19-94 (ASTM D 792)	g/ml	1,08	1,12
Hardness 25°C		IO-19-60 (ASTM D 2240)	Shore D/15	84	88
Glass transition (Tg)	7gg TA/RT	IO-19-69 (ASTM D 3418)	°C	55	61
	24hRT+15h 50°C		°C	70	76
	24hRT+15h 60°C		°C	79	85
			°C		
Maximum Tg		IO-19-69 (ASTM D 3418)	°C	94	100
Water absorption (24h RT)		IO-19-70 (ASTM D 570)	%	0,10	0,20
Water absorption (2h 100°C)		IO-19-70 (ASTM D 570)	%	0.60	0.70
Max recommended operating temperature		(***)	°C	90	
Flexural strength		IO-19-66 (ASTM D 790)	MN/m²	110	120
Maximum strain		IO-19-66 (ASTM D 790)	%	5,0	7,0
Strain at break		IO-19-66 (ASTM D 790)	%	6,0	8,0
Flexural elastic modulus		IO-19-66 (ASTM D 790)	MN/m²	3.200	3.600
Tensile strength		IO-19-63 (ASTM D 638)	MN/m²	67	75
Elongation at break		IO-19-63 (ASTM D 638)	%	6,0	8,0
Compressive strength		IO-19-73 (ASTM D 695)	MN/m²	91	103

IO-00-00 = ELANTAS Europe's test method. The corresponding international method is indicated whenever possible
nd = not determined na = not applicable RT = TA = laboratory room temperature (23±2°C)
Conversion units 1 mPas = 1 cPs 1MN/m 2 = 10 kg/cm2 = 1 MPa

(*) for larger quantities pot life is shorter and exothermic peak increases
(**) the brackets mean optionality
(***) The maximum operating temperature is given on the basis of laboratory information available being it function of the curing conditions used and
of the type of coupled materials. For further possible information see post-curing paragraph

A member of ⬤ **ALTANA**

PRODUCT INFORMATION   pag.3/4   ◐ **ELANTAS**

EC 157.1/W 152 LR

Instructions: Before use verify if components are perfectly transparent. Add the appropriate quantity of hardener to the resin, mix carefully. Avoid air trapping. If the mixing is carried on with dosing/mixing equipment deharation of the mixture is not necessary. On the contrary evaluate if it is necessary as function of vacuum applied during infusion.

Curing/Post-curing Post curing is always advisable for RT curing systems in order to stabilize the component and to reach the best properties. It is necessary when the component works at a high temperature. The rate of heating and the indicated post-curing time are referred to standard specimen size. Users should evaluate the best conditions of curing or post-curing depending on the component size and shape. For large size components decrease the thermal gradient and increase the post-curing time. In case of thin layer applications and composites, post cure on the jig. As general guide to minimize the risk of thermal deformations we suggest to carry on the post-curing in the following way:
- on mould: 24 h RT + 6 h 40°C + 6 h 50°C + 12 h 60°C
- out of the mould but on the jig: 7 days RT + 6 h 40°C + 6 h 50°C + 12 h 60°C

Storage: Unfilled epoxy resin and its amine based hardener can be stored for two years in the original sealed containers stored in a cool, dry place. The hardeners are moisture sensitive therefore it is good practice to close the container immediately after each use. The hardeners are moisture sensitive therefore it is good practice to close the container immediately after each use.

Handling precautions: Refer to the safety data sheet and comply with regulations relating to industrial health and waste disposal.

emission date:	May	2017
revision n° 00		

Manufactured: ELANTAS Europe S.r.l. Sito di Strada Antolini n°1, 43044 Collecchio (PR), Italy
www.elantas.com

A member of ◐ **ALTANA**

PRODUCT INFORMATION   pag.4/4   ⬤ **ELANTAS**

EC 157.1/W 152 LR

Reactivity profiles of the systems during mass reactions

(200ml system volume, resin/hardener mixing ratio 100:30 at 25°C in air)

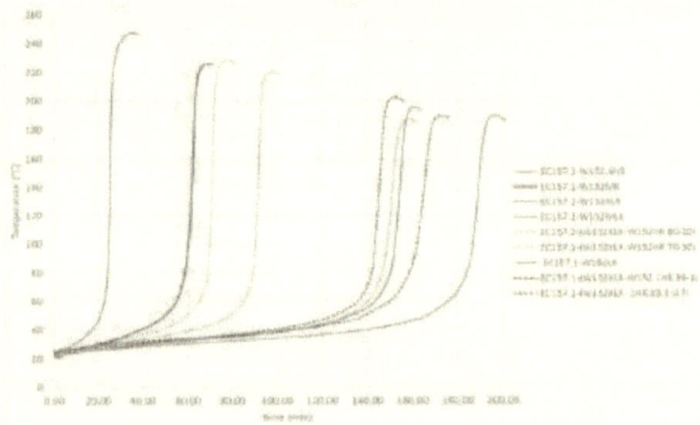

With HR label is identified the high reactivity hardener W152.1HR, generally suitable for small dimensions repairing or as reactivity modifier for other hardeners. The mixture of W152XLR with W152.1HR in different ratios allows to obtain intermediate reactivities. The systems are approved by DNV GL.

Processing times for the correct use of systems in vacuum infusion technology

	EC157.1/W152.1HR				EC157.1/W152VR				EC157.1/W152LR				EC157.1/W152XLR				EC157.1/W152XXLR			
Application temperature (°C)	15	20	25	30	15	20	25	30	15	20	25	30	15	20	25	30	15	20	25	30
Gel time (min)	[illegible]	[illegible]	[illegible]	[illegible]	[illegible]	[illegible]	[illegible]	[illegible]	[illegible]	[illegible]	[illegible]	[illegible]	[illegible]	[illegible]	[illegible]	[illegible]	[illegible]	[illegible]	[illegible]	[illegible]
Processing time before viscosity increase (min)	[illegible]	[illegible]	[illegible]	[illegible]	[illegible]	[illegible]	[illegible]	[illegible]	[illegible]	[illegible]	[illegible]	[illegible]	[illegible]	[illegible]	[illegible]	[illegible]	[illegible]	[illegible]	[illegible]	[illegible]
Demolding time (h)	[illegible]	[illegible]	[illegible]	[illegible]	[illegible]	[illegible]	[illegible]	[illegible]	[illegible]	[illegible]	[illegible]	[illegible]	[illegible]	[illegible]	[illegible]	[illegible]	[illegible]	[illegible]	[illegible]	[illegible]

N.B. The reported values are derived from lab tests and from the application experience. They must be considered indicative because they are related to the specific size and shape of the composite manufactures. Buyers and users should make their own assessments of our products under their own application conditions.

Interlaminar shear strength (ILSS) of laminates

(Unidirectional glass 600g/m² realized with infusion technology) –ASTM D2344

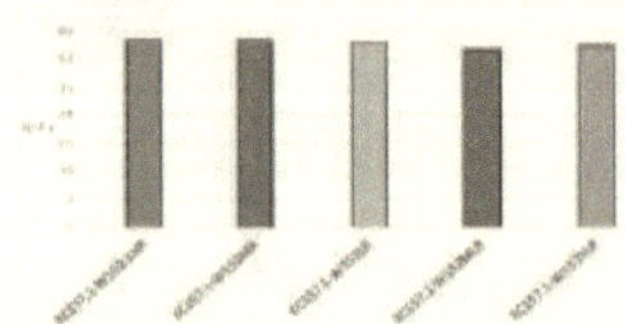

The composite laminates has been obtained by infusion of a 600gm² glass E tissue. From laminating of 5 mm of thickness cured at room temperature and stabilized at 50°C for 16 hrs were obtained specimens following ASTM D2344 code.

Chapter 3

Cucurbit[8]uril based crystal engineering

3.1 Introduction

As described in the previous chapters, Cucurbit[n]urils (CB[n]) serve as versatile host molecules capable of guest binding with good selectivity and offer possibilities for molecular recognition and the construction of molecular machines and rotaxanes.[1, 2] They also provide a platform for the development of supramolecular organic frameworks and the formation of materials with inherent porosity.[3, 4] In this context, the fundamental study of the stochiometry and geometry of CB[8] and CB[8] complexes in the solid state is pivotal. Understanding the different modes of interaction between CB[n] in the solid state may allow us to design and synthesise 1D, 2D and 3D crystalline assemblies that could be used for various applications such as gas adsorption, separation and energy storage.[5–8] This chapter explores different crystalline assemblies of CB[8] isolated by the use of chaperone molecules and metal ions. Two porous metal free nanotubular structures formed by chaperone molecules and two metal coordinated CB[8] structures are discussed.

3.1.1   Cucurbit[8]uril packing in solid state

Ripmeester and coworkers published a comprehensive solid state study of CB[5] to CB[8] where they used a combination of techniques to identify a number of general features to reveal interesting properties of these macrocycles, including their solubility, crystallinity, stability and aggregation.[9, 10] In the solid state, cucurbiturils show a tendency to interact through weak C−H⋯O bonds involving the methene and methine protons on their outer ring and the carbonyl oxygens of a neighbouring CB (Figure 3.1).[11, 12]

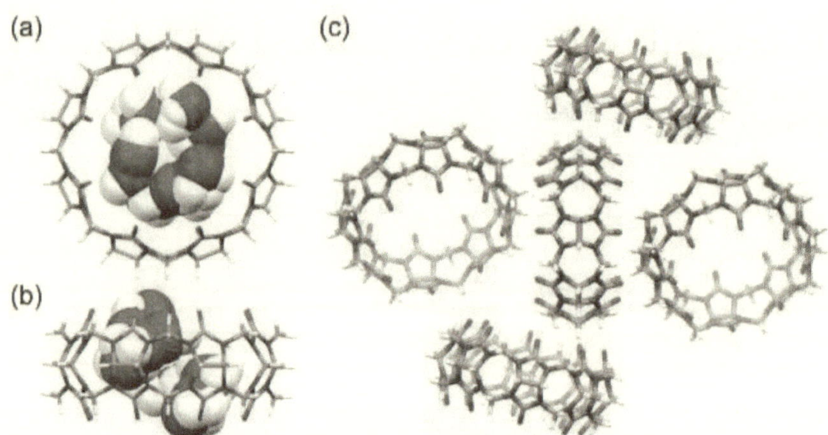

Figure 3.1: (a) Inclusion of disordered water in CB[8]. (b) Self-closing interaction in which CB[8] is not engaged in other competitive interactions such as complexation. Adapted from reference [13].

During the synthesis of CB[8], purification is most effectively carried out by re-crystallisation from hot concentrated HCl leading to the formation of large colourless crystals. CB[8] crystallises in the tetragonal space group $I4_1/a$, yielding the most common packing of CB[8] formed in the absence of competing interactions. The structure reported by Ripmeester and coworkers contained 4 HCl and 15 water molecules per CB[8]. Within the cavity, disordered water was observed as shown in Figure 3.1 (a) and (b). Here the driving force of the assembly is the large number of C−H⋯O between the carbonyl oxygens and the methene and methine protons, essentially leading to a self closing of the cavity. Figure 3.1(c) shows the crystal packing where water and chloride ions have been removed for clarity. In these assemblies, adjacent macrocycles are positioned in an alternating arrangement, almost perpendicular to each other.

Ripmeester and coworkers were able to model all the hydrogens in this structure and show the precise role of hydrogen bonding in the assembly. It was observed that there were a relatively high number of weak H-bonds, where 3-5 oxygens participated in 9 inter cucurbituril H-bonds. The rest of the carbonyl oxygens participated in classical water mediated H-bonds. The CB[8] molecules further stack to form a 3 dimensional network. Figure 3.2 shows the consecutive layers I to IV that are present along the b axis. In the channels formed by this network chloride ions and water are present that interact with the outer walls of the CB[8].

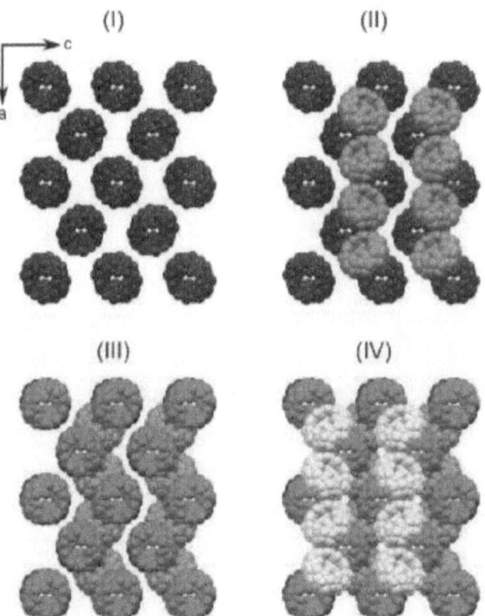

Figure 3.2: Packing of CB[8] showing different layers I,II,III and IV. Adapted from reference [13].

It is clear that it is a combination of the weak interactions, specifically the multiplicity of the inter-cucurbituril weak H-bonds that stabilise or at least direct the formation of these cucurbituril crystals in water. Especially in the larger CB[8], the high number of these weak interactions per macrocycle allows these crystals to show quite remarkable thermal stability.[14] However, these forces are still weak and can be easily overruled when other interactions are active in solution during crystal growth. Examples of other interactions are (i) the presence of cations that interact with the ureido carbonyls forming stronger ion dipole interactions overruling any $C-H\cdots O$, thereby leading to the formation of different crystalline phases of CB[n]s and (ii) the interaction of a CB[n] with guest molecules in solution.

Upon synthesis of CB[8], the crystal structure of the obtained product was found to be isostructural to those reported in the literature, with Cl^- ions from recrystallisation in concentrated HCl.[13, 15].

3.1.2   Cucurbit[n]uril based metal coordination

Recently, there has been increasing interest in the area of metal organic frameworks due to their widespread applications in a broad range of fields.[16] Due to the presence of electronegative ureido carbonyls on their portals, in addition to binding organic molecules, CB[n]'s are promising ligands for coordination with metal ions.[17] The direct coordination of metal ions to the carbonyl oxygen atoms of the glycoluril units can lead to the formation of very interesting structures with a high scope for applications in new materials. The different homologues of CB[n] show very different coordination tendency to different metal ion, with most of the work done on the smaller homologues from n = 5-7. In addition, these assemblies are heavily dependant on various other environmental factors such as counter ions, solvent, pH and metal ligand ratio.

The most common structural motif of metal - CB[n] coordination compounds is the molecular capsule or molecular bowl. Some of the first studies of CB[n] coordination were reports of the formation of a CB[6] molecular 'capsule' by Kim and coworkers.[18] In this capsule, an inclusion complex of a molecule of tetrahydrofuran (THF) within CB[6] was capped by Na^+ ions, as shown in Figure 3.3. The release and resealing of the lid could be controlled by the complexation and decomplexation of the sodium ion by varying the pH, since the metal was released in strongly acidic conditions. Similar molecular capsules and bowls were also constructed using the smaller CB[5] and a number of metals including alkali K^+, alkaline earth Ba^{2+}, transition Cd^{2+} and lanthanide La^{3+} metal ions.[19, 20]

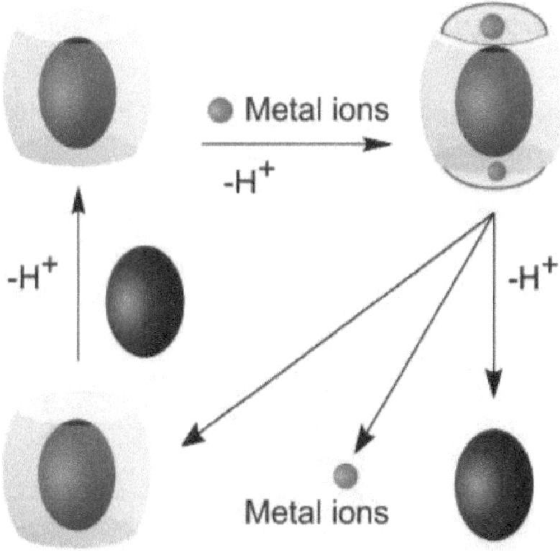

Figure 3.3: Schematic representation of CB[6] based molecular capsule reported in Reference [18]

Coordination complexes of CB[n], n≥7 with metal ions

The coordination of metals to higher homologues of CB[n] is significantly more difficult than CB[5] and CB[6], with only a few complexes reported in the literature. This is thought to be due to the larger portal size, leading to a wider distribution of the electronegative carbonyl oxygens. Since the portals cannot be fully capped, a few metal cations or complexes have been found either forming inclusion complexes or in the lattice spaces outside the host. Thuéry reported the synthesis of novel CB[7] based 2D frameworks through the coordination of uranyl ions to the carbonyl portals.[21] Under solvothermal synthesis in the presence of a 10 fold excess of $UO_2(NO_3)_2$, CB[7] arranged into tubular polymers forming a CB[7] based metal organic framework. So far, very few typical CB[8] based metal coordination complexes have been successfully characterised by X-ray crystallography. The first known examples of metal coordinated CB[8] are shown in Figure 3.4. The first structure was a complex of CB[8] and a polynuclear strontium aqua complex where two bidentate Sr^{2+} ions are coordinated to each portal of such that each CB[8] binds 4 cations.[22] In the second structure CB[8] acts as a tetradentate ligand, bidentate at each portal, binding 2 $[Bi(NO_3)(H_2O)_5]^{2+}$ cations per host.[23] Following this, Zhu and coworkers described a novel 3D framework made from four CB[8] molecules through coordination of caesium ions to the portal oxygens of the CB[5] induced by the guest p-hydroxybenzoic acid.[24, 25]

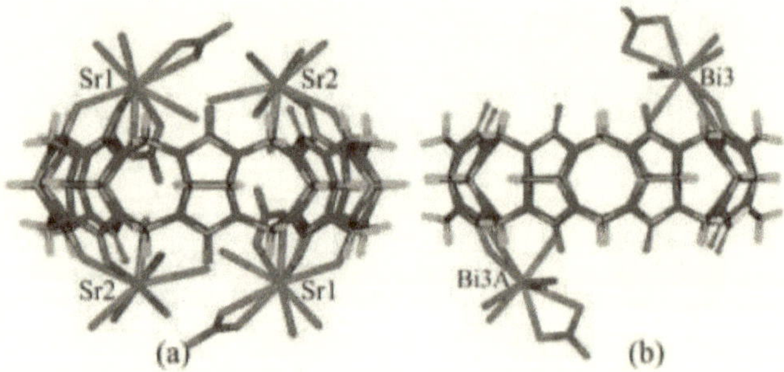

Figure 3.4: CB[8] metal coordination complexes.[17]

3.1.3   Columnar arrangement of CB[8]

Ripmeester and coworkers observed two new crystalline phases of CB[6] and CB[8] during the synthesis and purification of these homologues.[10] The CB[8] structure was observed upon crystallisation from nitric acid. The interesting feature of these structures was that the macrocycles were stacked in a perfectly tubular arrangement in one dimension as shown in Figure 3.5. In the case of the CB[8] structure, the crystal shows two distinct channels, the first formed by the cavities of the macrocycles piled one on top of the other, and the second formed by the voids between the CB[8]. This structure also showed the highest water content reported in any CB[n], with 60 per macrocycle. However this resulted in an unstable structure that quickly showed cracks when exposed to air. Nevertheless, such tubular architecture especially of CB[8] is uncommon, due to its self closing tendency.

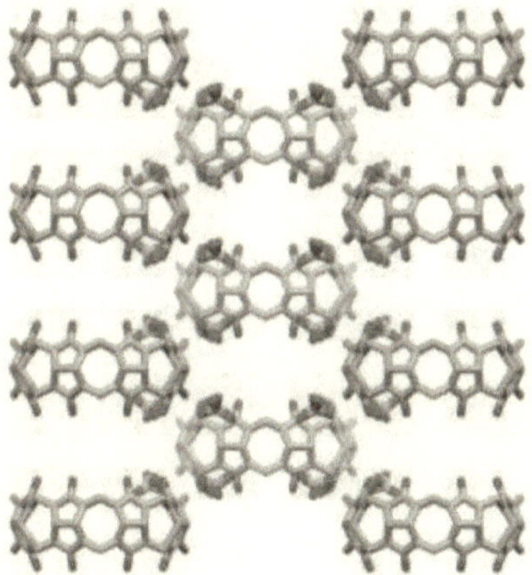

Figure 3.5: Higly symmetric CB[8] assembly (Figure adapted from Reference [10]).

3.2 Chaperone assisted nanotubular framework

3.2.1 Molecular chaperones

Li and coworkers reported a new example of a small organic molecule acting as a 'chaperone molecule' for guiding the formation of a porous assembly in CB[8].[26] The key point of this report was the formation of a 1:1 complex with the chaperone molecule in solution. At conditions of high temperature and pressure, it was observed that the solution of this complex in 1:1 water:DMF caused the crystallisation of pure CB[8] with a honeycomb like structure, leaving the excluded guest in solution. DMF also participated in stabilising the 3D supramolecular structure.

Following this, Cao and coworkers reported the formation of a crystalline nanotubular framework formed by CB[8] by a template effect caused by a bipyrimidine derivative functionalised with C_4H_8COOH that forms a 1:1 complex with CB[8] as a chaperone molecule.[27] This framework was observed to show high thermal stability and permanent porosity. It was observed that the assembly showed highly selective adsorption for CO_2 over N_2, H_2 and CH_4. The formation of the assembly was proposed to occur by the formation of the 1:1 complex and the re-dissolution of the chaperone molecule in solution during crystallisation.

The pivotal step for the formation of both the above CB[8] was the association and dissociation of the host CB[8] and the respective guests. The guest molecules therefore showed a chaperoning effect towards the formation of these assemblies. The term molecular chaperone was first coined to describe a diverse family of cellular proteins which assist in the correct folding and sometimes assembly of other polypeptides but which are not then components of the resulting final structure while they perform their normal biological function.[28] Figure 3.6 shows a general representation of molecular chaperones.

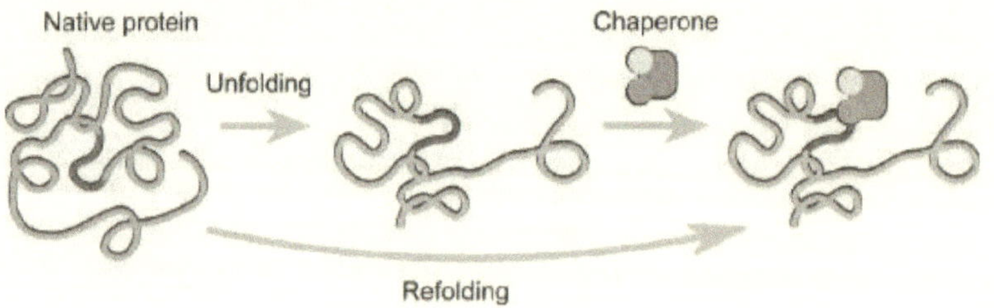

Figure 3.6: Schematic representation of molecular chaperones in protein folding.

Laskey et al during the investigating of the formation of nucleosomes[29] observed that once the interactions between the DNA and the histone core of the nucleosome was disrupted, the addition of the two components at physiological concentrations did not cause their self-assembly in the correct orientation and instead resulted in the formation of insoluble aggregates. However, the presence of an acidic protein called nucleoplasmin in the solution resulted in the formation of the correct assembly. Importantly the acidic protein is not a component of the final nucleosome, and merely provided a chaperone effect. Since then, chaperone based mechanisms have been used in complex biological systems, in the directed folding of proteins upon synthesis and in the prevention of diseases caused by misfolding of proteins and so on. There have also been a number of studies on small molecules designed to be able to provide a chaperone effect, bind proteins and restore their original conformations.

The chaperone assisted assembly strategy has a high potential for the design and synthesis of artificial supramolecular systems. They can be used to provide synthetic strategies for assemblies that would not be otherwise formed. Some examples of chaperone effect are the work of Fujita who showed the chaperoning effect in the formation of long coordination nanotubes.[30–33]

3.2.2   CB[8] nanotubular framework 1 (CBT-1)

During our early investigation of suitable probes for use in the damage reporting polymers described in Chapter 2, one of the complexes studied was that of CB[8] encapsulating a butyl viologen and pyrene derivative within the cavity. The formation of the ternary complex was thoroughly investigated in solution by UV-Vis, fluorescence and NMR spectroscopy, following which we made several attempts to grow single crystals of this ternary complex in order to understand the orientation of the guest molecules within the host cavity. In the presence of butyl viologen, stable octahedral crystals were repeatedly obtained. The structure was not fully solved due to a high amount of disorder, but it was clearly observed that the CB[8] molecules in the crystal were packed very differently from the traditional interpenetrated packing that is generally observed as shown. The tubular assembly is shown in Figure 3.7.

Crystallographic analysis hinted at the absence of any guest molecule in the cavity of CB[8]. The tubular framework of the CB[8] interested us and we found that this structure was also obtained when the butyl chain on the viologen was replaced with a shorter methyl substituent. We therefore hypothesised that the packing of the CB[8] was due to a template effect by the guest molecules.

In order to test this hypothesis we attempted the synthesis of tubular CB[8] by re-crystallisation in the presence of viologens with varying length of side chains as molecular chaperones. 0.0025 mmol of CB[8] and methyl viologen diiodide were dissolved in 2 mL of water and heated to 60 °C for 5 h. The solution was filtered and kept in a vial for slow evaporation. Large colourless crystals were observed after only a few hours at room temperature.

The crystal structure of **CBT-1** was determined via X-ray diffraction data

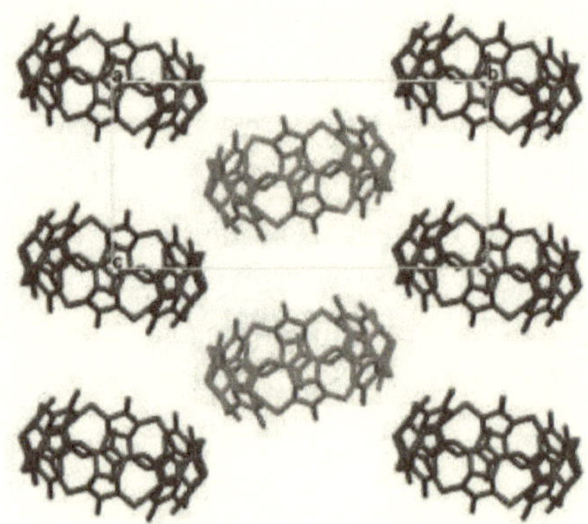

Figure 3.7: Tubular structure of CBT-1.

on these single crystals. The crystal structure indeed showed the presence of a three dimensional framework of CB[8]. The mechanism for the formation of this assembly is thought to proceed as shown in Figure 3.8. Methyl viologen is highly water soluble, while CB[8] is extremely sparsely soluble in water with a solubility ≤ 0.1 mM. Since the cavity of CB[8] is too large for the guest, and due to the inability of this particular guest to form any interactions with neighbouring CB[8], the molecule seems to prefer during the crystallisation process to escape the cavity and return to solution. Upon the publication of the report by Cao and coworkers we realised that they had observed a similar templating effect with a differently substituted viologen derivative, corroborating our hypothesis.[27]

CBT-1 crystallises in the orthorhombic space group *Pccn*. The asymmetric unit consists of half CB[8] macrocycle, one chloride anion, half iodide anion and 11 lattice water molecules and is shown in the experimental section. To balance the negative charge brought by the anions, the presence of one and a half protonated water molecules in the asymmetric unit must be supposed, even if the position of the protons could not be identified in the difference Fourier map. The crystal structure is as usual consolidated by a network of H-bonds involving the lattice water molecules, the anions, and the macrocycles, and by van der Waals interactions. The assembly is a 3D nanotubular framework, with

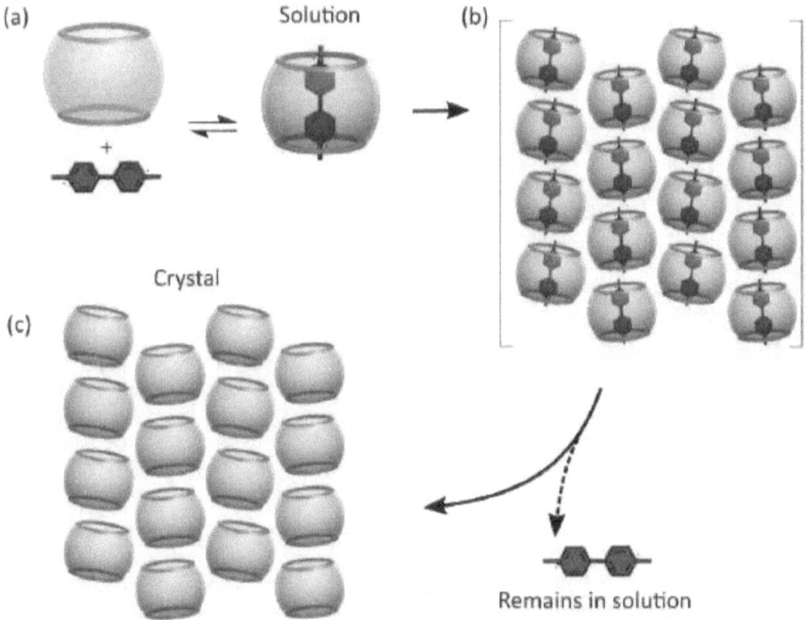

Figure 3.8: Schematic representation of proposed chaperone based mechanism for formation of Framework I.

intrinsic 1D nanotubes formed by the inner-cavities of CB[8] and additional 1D extrinsic channels formed by the other surfaces of CB[8] that contain iodide ions. The CB[8] molecules are tilted at an angle to each other as shown in Figure 3.9.

Figure 3.10 shows the two distinct layers in the packing of **CBT-1**. Repeating the crystallisation trials by varying the chain length of the viologen from C_1 to C_4 resulted in the formation of the same structure. We also repeated the experiment by increasing the chain length drastically to C_{11}, since this large aliphatic chain should not allow the individual host guest systems in solutions to pack together and should not provide the same template effect. As expected, this framework was not formed, however an interesting structure of one molecule of MV-C_{11} encapsulated head and tail by 2 molecules of CB[8] was observed. This structure is described in detail in Chapter 4.

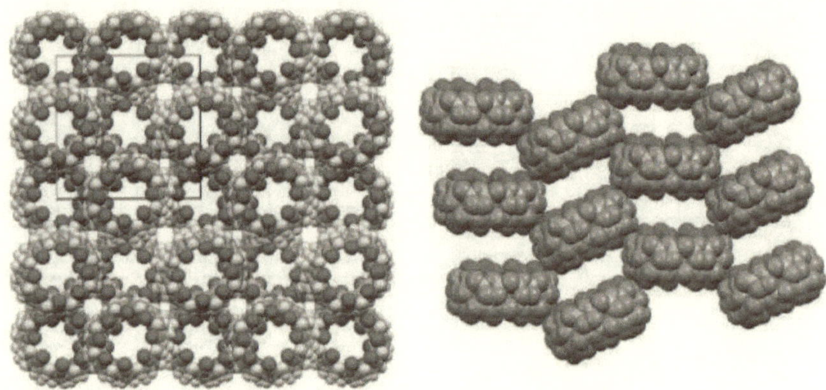

Figure 3.9: Packing of the CB[8]s viewed along the *c*- and *b*-axis direction of the unit cell (left and right, respectively). The lattice water molecules and the anions have been omitted for clarity.

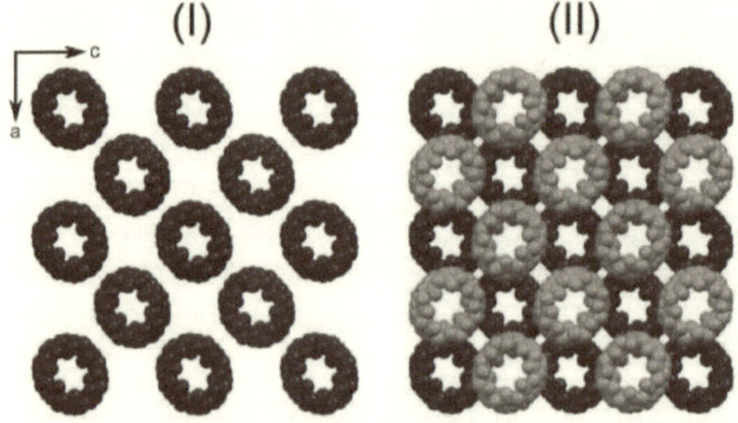

Figure 3.10: Packing of channelled CB[8] showing different layers I and II.

A few crystals were taken out, washed and dried and were found to be stable in air. They were then dissolved in DMSO-d$_6$. NMR analysis confirmed the absence of guest molecules within the crystal structure, showing the presence of the protons of CB[8]. Since the crystals were stable over large periods of time, we attempted to expose them in different guest molecules to assess the accessibility of the cavity. The crystals were exposed to I_2 vapour, upon which they rapidly changed their colour to a deep brown. The crystalline framework however was unchanged. Upon soaking the crystals in a solution of pyrene-imidazolium bromide over 24 h, there was a slight colour change in the crystal. However upon examination by single crystal X-ray diffraction it was found that the basic structure of the assembly was unchanged. It was interesting to note, however that the iodide counter ion had been replaced by the bromide counter ion of the guest found in solution. This was an interesting result as it shows adaptive nature of the pores of the framework to encapsulate anionic guests from solution. This could have potential applications as a soft supramolecular material that could be used to effectively remove potentially dangerous anions from solution such as $[\mathrm{TcO_4}]^-/[\mathrm{ReO_4}]^-$.[34] Studies to further understand and ascertain the selectivity of anion exchange are underway.

It is also worth noting that even in the presence of additional CB[8] guests such as pyrene-imidazole, the presence of $\mathrm{MV-C_2}$ or $\mathrm{MV-C_4}$ always resulted in the formation of pure CB[8] in the form of CBT-1. This method could also be used to effectively regenerate pure CB[8] from a mixture in solution.

3.2.3   CB[8] nanotubular framework 2 (CBT-2)

When the above system was recrystallised in 6M HCl, a different nanotubular framework was obtained with perfectly parallel CB[8] columns. The asymmetric unit is shown in the experimental section. **CBT-2** crystallizes in the monoclinic space group $C2/m$ and its asymmetric unit is shown in the experimental section. To balance the negative charge brought by the anions, the presence of two protonated water molecules in the asymmetric unit must be supposed, even if the position of the protons could not be identified in the difference Fourier map. This is also supported by the crystallisation of the compound from water acidified by HCl. The crystal structure is consolidated by a network of H-bonds involving the lattice water molecules and the chloride anions, as well as the macrocycles. It is noteworthy that the CB[8]s are packed in a stacking fashion giving rise to channels along the b-axis direction of the unit cell (see Figure 3.11), with an average diameter of roughly 10 Å.

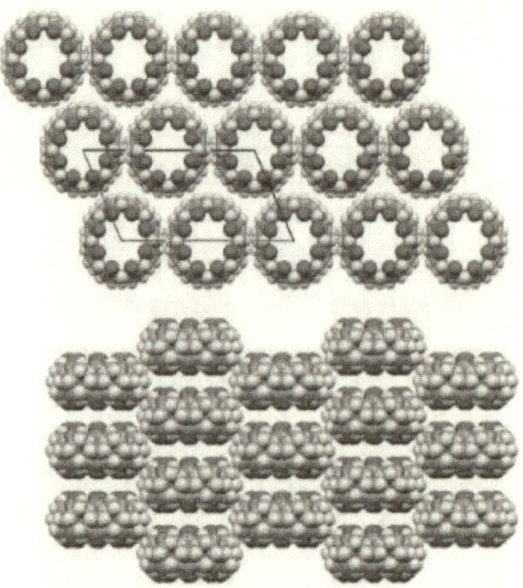

Figure 3.11: Packing of CBT-2

3.3   CB[8] metal coordinated tubular assembly (CBT-3)

As mentioned earlier materials with tubular architectures have wide applications in the fields of nanotechnology, ion sensors, and so on.[35–38] One disadvantage of the columnar packing of CB[n] reported so far, including the CB[8] reported earlier in this chapter by the chaperone based approach, is that this supramolecular framework is not stable enough to maintain the porous structure. Additionally, a fair amount of water is involved in maintaining the crystal structure through hydrogen bonding and the removal of all the water molecules from the framework in order to completely free the pores for adsorption is not possible without disrupting the framework. Coordination polymers of metal ions and CB[n] molecules through direct coordination could lead to the formation of more stable porous materials to overcome this limitation. Though the carbonyl oxygens of the CB[n] portals can coordinate to metal ions, the one dimensional assembly of tubular polymers through CB[n] and direct coordination of metal ions is still uncommon. Kim and coworkers showed the first such coordination polymer formed by the coordination of the portal oxygens of CB[6] to rubidium in an alternating manner.[39] This assembly formed a one-dimensional tubular polymer, where the polymer chains were further arranged to form a honeycomb structure with large linear channels parallel to the chains. The coordination of CB[6] to alkali metal ions, specifically sodium ions[40] and potassium ions[41] to form similar structures has also been reported.

Since direct metal coordination of CB[8] is already rare, it is also evident that such tubular assemblies of CB[8] are not easily formed. We observed such a structure by inducing the formation of a perfectly columnar assembly of CB[8] formed through direct coordination of a copper ion upon hydrothermal treatment of CB[8] and copper nitrate in the presence of 4,4' azopyridine at 100 °C for 24 h (**CBT-3**). Each CB[8] portal is coordinated to two copper ions through opposite carbonyl oxygens. These copper ions further coordinate the carbonyl oxygens of another CB[8] in a linear fashion as shown in Figure 3.12 leading to the formation of a tubular structure. Such a perfectly parallel tubular framework of CB[8] through direct metal ion coordination has not yet been reported. This structure is currently under refinement.

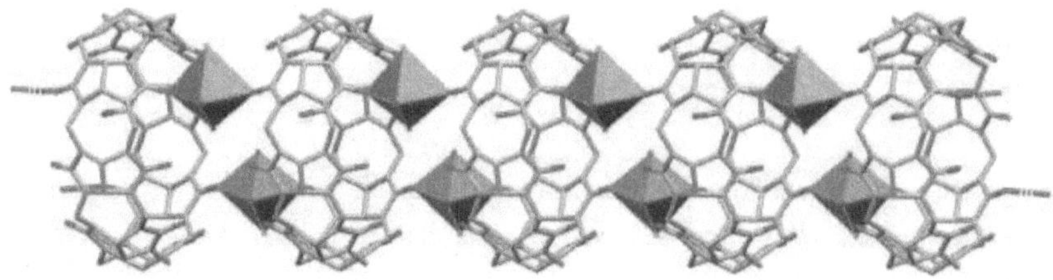

Figure 3.12: Tubular assembly formed by coordination of copper ions to portal oxygens of adjacent CB[8]s

The linear channels in **CBT-3** are arranged side by side as shown in Figure 3.13. Each column has a self-complementary curvature with alternating 'bumps' and 'hollows'. In the solid-state, the next column is offset by one-half repeating unit along the column direction to maximise the interactions between the two columns by fitting the bumps into the hollows of the neighbouring columns.[41]

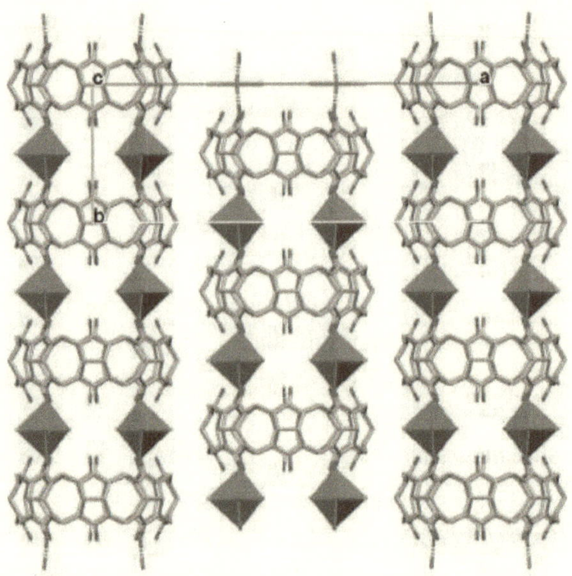

Figure 3.13: Tubular copper coordinated CB[8] channels stacked side by side

3.4 Assembly of CB[8] in the presence of $[ZnCl_4]^{2-}$

Nearly all CB[n] research has focused on using the carbonyl portals and the cavity to drive the formation of supramolecular assemblies due to the strong charge–dipole and hydrogen bonding interactions, as well as hydrophobic and hydrophilic effect derived from the negative portals and their rigid cavities. A recent study revealed that other weak noncovalent interactions could also be involved in driving such assemblies, other than hydrogen bonding and $\pi\cdots\pi$ stacking, for instance C–H$\cdots\pi$ and ion–dipole interactions derived from the electrostatically positive outer surface of the CB[n]s (Figure 3.14). These interactions, dubbed as outer surface interactions could also be driving forces in the formation of novel CB[n]-based supramolecular architectures and could also drive the formation of assemblies that could not form otherwise.[16, 40]

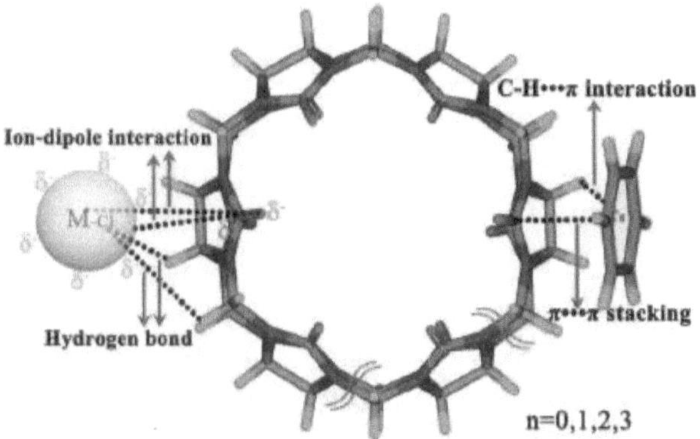

Figure 3.14: Possible outer-surface interactions between CB[n] and Inorganic Molecules between CB[n] and Aromatic Molecules. (Adapted from reference [17])

In recent years, numerous research results have revealed that when a third species is introduced into the metal CB[8] system, the coordination as well as the supramolecular assemblies can be dramatically different from those in the absence of the third species. In particular certain transition metal ions have recently been introduced into metal–cucurbit[n]uril systems in an HCl solution, resulting in the formation of supramolecular assemblies with very different arrangements from those formed without these inorganic structural inducers.[42] It was observed that transitional metal ion complexes such as $[CdCl_4]^{2-}$ and $[ZnCl_4]^{2-}$ with CB[n] could form a honeycomb structure.[43] Additionally, In the presence of lanthanide ions, zigzag and linear coordination polymers were formed with the CB[n] arranged within the honeycomb structure, so that each single coordination polymer is surrounded by the transition metal counter ion as shown in Figure 3.15.

Tao and coworkers proposed that the were two main driving forces of these outer-surface interactions. The first was the unusual hydrogen bonding of methine or methylene groups on the outer surface of CB[8] molecules with $[MtransCl_4]^{2-}$ anions or portal carbonyl oxygens of CB[n] molecules.[16] The second was the ion–dipole interaction of $[MtransCl_4]^{2-}$ anions or portal carbonyl oxygens with portal carbonyl carbons of the adjacent CB[n] and of the

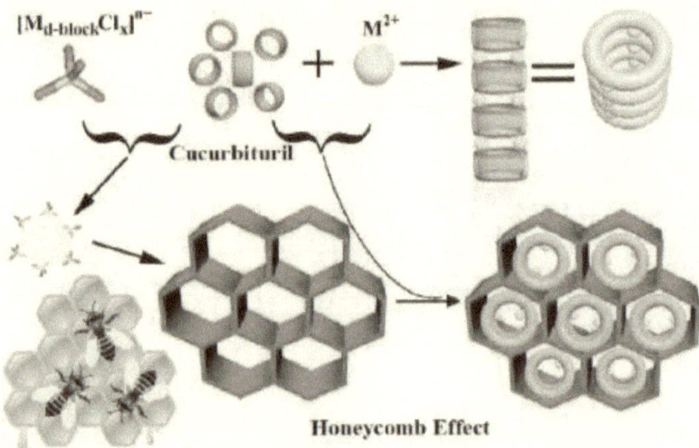

Figure 3.15: Diagram of the honeycomb effect promoting coordination of CB[n] (Adapted from reference [42])

satellite $[MtransCl_4]^{2-}$ anions. Overall, there is an interaction of the positive electrostatic potential on the outer surface of CB[n] with the $[MtransCl_4]^{2-}$ anions.

These studies showed that there is a preference for outer surface interaction over direct metal coordination for CB[n]s. Detailed studies of this phenomenon in CB[8] were carried out by Ji *et al*, by studying the assemblies of CB[8] in the presence of tetrachloride transition metal ions ($[MtransCl_4]^{2-}$, Mtrans = Cd, Zn, Ni).[44, 45] This assembly showed a parallelogram channel constructed of CB[8] molecules and $[CdCl_4]^{2-}$ anions, and introduced a new strategy to synthesise CB[8]-based porous materials using transition metal chloride anions as structure inducers. In their report they also suggested that these porous materials could also capture organic molecules within the cavity of the CB[8].

Though perylene diimide (PDI) complexation in solution has been well characterised, growing crystals of CB[8] with large guests such as PDI has proved difficult and no crystal structures of PDI and CB[8] have been reported so far. We attempted to grow crystals of CB[8] with a PDI guest molecule in the presence of transition metal anions of the type reported above, in an attempt to lend rigidity to the framework. However during the crystallisation process, very long rectangular crystals were observed, upon which there was a layer of the guest. When these crystals were examined by single crystal X-ray diffraction, it was observed that there was no inclusion of the guest, however a three dimensional framework of CB[8] was observed in the presence of $[ZnCl_4]^{2-}$ anions. The cell is shown in Figure 3.16 and comprises CB[8] molecules displaying two different orientations. One set of CB[8]s are arranged in channels between which another set of CB[8]s are located at a tilted angle. The $[ZnCl_4]^{2-}$ occupy the voids among the CB[8] macrocycles. Both the macrocycles are empty, with the exception of some water molecules located inside the cavity. The crystal structure is consolidated by H-bonds involving the water molecules, the anions and both the CB[8]s, and by van der Waals forces. The packing is shown in Figure 3.16, highlighting the reciprocal position of the macrocycles which are slightly inclined with respect to each other.

Figure 3.16: Unit cell of the 3D CB[8] assembly

The packing of the CB[8] - $[ZnCl_4]^{2-}$ assembly is shown in Figure 3.17. The assembly is different from that observed in the case of Tao.[44] It is currently unclear whether the inclusion of the perylene in the CB[8] in solution which has been well established promoted the formation of this assembly much like in the case of the molecular chaperone in the first assembly described in this chapter. The crystallisation of CB[8] with $[ZnCl_4]^{2-}$ in the absence of the guest

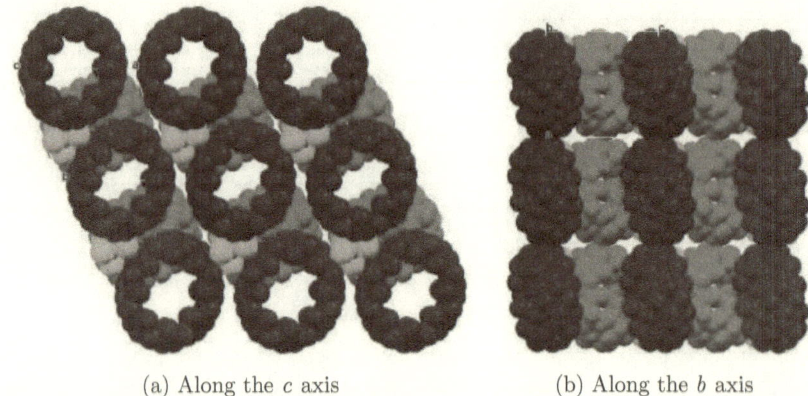

(a) Along the c axis   (b) Along the b axis

Figure 3.17: Packing of the CB[8] $[ZnCl_4]^{2-}$ framework. The two types of CB[8]s showing different orientations are shown in blue and red. H atoms and lattice water molecules have been omitted for clarity.

is currently underway to shed more light on this issue.

3.5 Conclusions

In this chapter, we have shown four tubular assemblies of CB[8]. **CBT-1** and **CBT-2** formed by a chaperone based strategy show promising potential for efficient creation of supramolecular systems, and should provide a more facile way to achieve new assemblies inaccessible using the conventional direct methods. We also report a tubular assembly by the direct coordination of a metal ion to CB[8] carbonyl oxygens. The direct coordination of metal ions to CB[n] generally results in assemblies which are more stable to the removal of solvent molecules. The assembly of metal-CB[8] systems formed through direct coordination with tubular architectures have potential application as sensors, catalysts and can also provide a porous framework for the adsorption of guests too large for the smaller CB[n]s.

It is also evident from the observations that due to the differential solubility of the CB[8] and its complexes as well as the intrinsic tendency of the CB[8] to form well packed compact structures proves a challenge to the stabilisation of host-guest complexes in the solid state. The guest molecules often tend to slip out during crystallisation unless additional noncovalent interactions can stabilise the framework. In the next chapter, we show how we have addressed this issue by growing assemblies of CB[8] with ditopic guests that are able to form polymeric frameworks by H-bonding and π stacking.

3.6 Experimental section

Synthesis

Synthesis of CB[8]

Glycoluril(**1**) was synthesised from urea and glyoxal. To a flask, urea (25 g, 414 mmol, a 40% aqueous solution of glyoxal (25 mL, 172 mmol) and 80 mL water was added, followed by the dropwise addition 2 mL of conc. sulphuric acid during stirring. The reaction was then stirred at 90 °C for 5 min. A white solid was formed soon after and the reaction mixture was heated for another 12 h. The solution was then cooled to room temperature and a solution of 50% NaOH was slowly added to bring the pH up to 14. The solution was then cooled and the precipitate was filtered out to give the pure product.

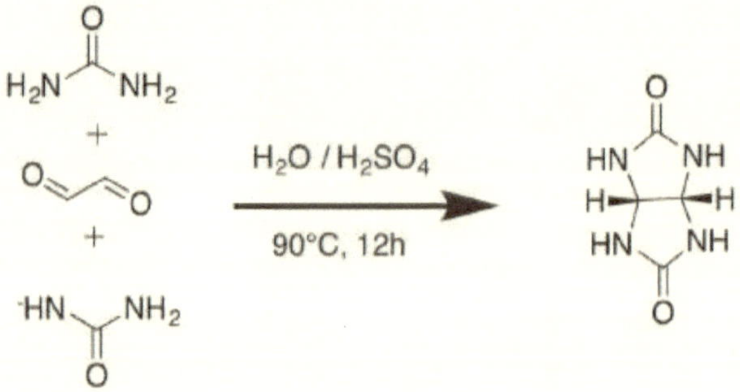

Scheme 3.1: Synthetic scheme for glycoluril(**1**)

CB[8] was synthesised and purified by a procedure based on methods of Kim and Day.[15] To glycoluril(1) (10 g, 74 mmol) in a round bottomed, a cooled solution of 37% HCl solution (15 mL) was added. Paraformaldehyde (4.2 g, 141 mmol) was slowly added to this mixture during rigorous stirring. The viscous solution was stirred, and within half an hour set as a gel. The reaction mixture was then heated to 100 °C, during which time the gel proceeded to melt. After stirring at 100 °C for around 18 h the stirring was stopped and the reaction was cooled to room temperature and left for 4 h wherein a precipitate as observed. This precipitate was filtered off and the filtrate was subsequently evaporated to give a sticky solid.

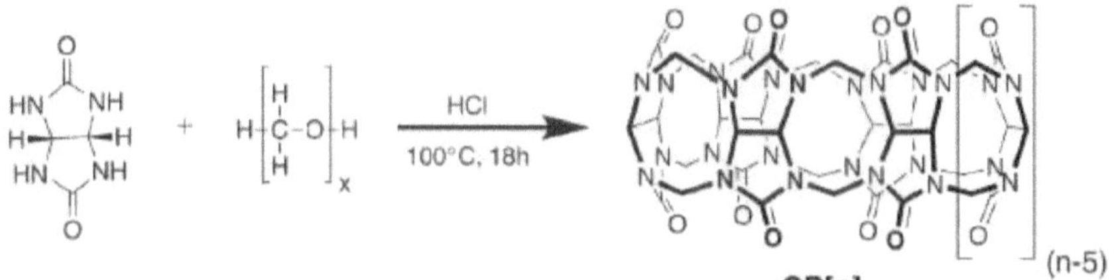

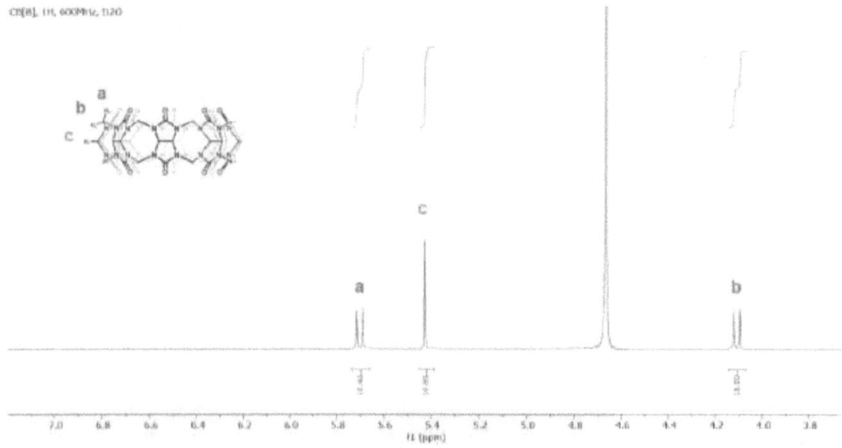

Scheme 3.2: Synthetic scheme for CB[8]

In order to remove the other homologues of CB[n], a 40% formic acid solution was added to this solid and the mixture was stirred for 1 h. CB[8] is
slightly soluble in this mixture, so the amount of formic acid used should be the
minimal required to sufficiently solubilise the other components of the mixture
while losing as less CB[8] as possible. After digesting the mixture with formic
acid twice, the remaining material was then recrystallised from hot concentrated HCl (37%) over a few hours to give large crystals. These crystals were
filtered and briefly washed with a small amount of water.

^{1}H NMR δ = 4.28 (d, 2H, J = 15.2 Hz, CH$_2$), 5.60 (s, 2H, CH), 5.89 (d,
2H, J = 15.2 Hz, CH$_2$)

Figure 3.18: ^{1}H-NMR of purified CB[8] in D$_2$O, 600 MHz

The purity of CB[8] is tested by ITC titration with adamantyl amine (ADA) which is known to form 1:1 complexes with CB[8] with high affinity. Stock solutions of CB[8] at 0.1 mM and adamantyl amine at 1 mM were prepared in 50 mM Sodium acetate buffer at pH = 4.75. The binding curve is shown in Figure 3.19. The effective purity of the synthesised CB[8] was therefore 73% which was used for all further calculations.

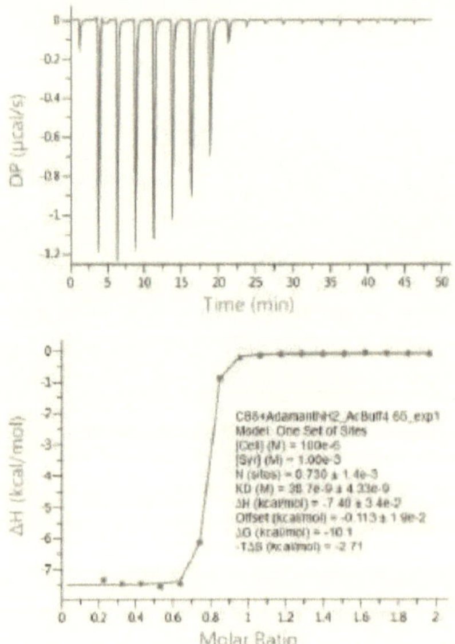

Figure 3.19: ITC binding curve for ADA and CB[8].

Synthesis of Methyl Viologen

4,4'-Bipyridine (0.5 g, 3.2 mmol) and excess methyl iodide (15 mmol) were dissolved in 50 mL anhydrous acetonitrile. The mixture was heated to 80 °C and stirred overnight. After cooling down to room temperature, the solvent was evaporated under reduced pressure. The residue was dried in vacuum to remove any residual iodomethane, and the resulting product was obtained as a red powder. (1.3 g. 95%). 1H NMR (400 MHz, D_2O): δ = 4.51 (s, 6H), 8.53 (d, J=6.26 Hz, 4H), 9.06 ppm (d, J=6.32 Hz, 4H)

Synthesis of nanotubular assembly CBT-1

The first crystal of CBT-1 was found during the crystallisation attempt of methyl viologen, pyrene and CB[8] in water by solvothermal synthesis. The experiment was then repeated without the third pyrene guest. Methyl viologen diiodide (1 mg, 0.0025 mmol and CB[8] (4.5 mg, 0.0025 mmol) were placed in a glass vial with 1 mL of millipore water. The vial was capped and heated in an oven at 60 °C overnight, filtered and allowed to stand. Within a few hours, clear colourless crystals were observed that grew progressively larger with time. They were harvested for collection after 1 week.

Synthesis of nanotubular assembly CBT-2

Methyl viologen diiodide (1 mg, 0.0025 mmol), CB[8] (4.5 mg, 0.0025 mmol) and pyrene imidazole derivative (1 mg, 0.0025 mmol were dissolved in a glass vial with 0.5 mL of 6M HCl. The vial was capped and allowed to stand. Long rectangular crystals grew over the course of months. The crystals were pale yellow.

X-ray fluorescence element analysis

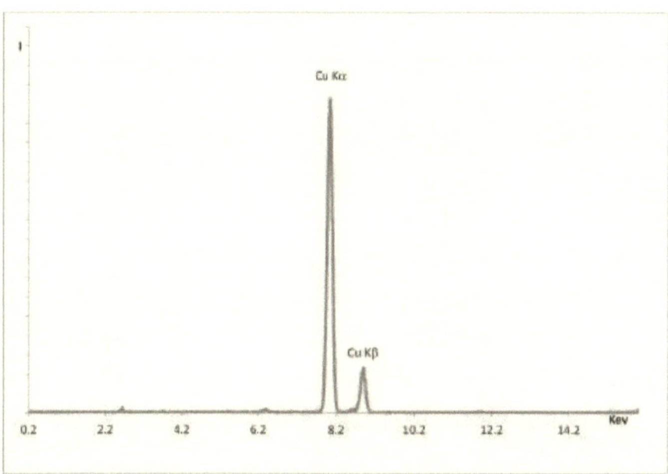

Figure 3.20: Confirmation of presence of Copper in **CBT-3**.

X-ray crystallographic section for Assembly 1 and 2

The crystal structures of compounds CBT-1 and CBT-2 were determined by X-ray diffraction methods. Crystal data and experimental details for data collection and structure refinement are reported in Table 3.1.

For CBT-1, intensity data and cell parameters were recorded at 100(2) K at the ELETTRA Synchrotron Light Source (CNR Trieste, strada statale 14, Area Science Park, 34149, Basovizza, Trieste, Italy). The raw frame data were processed using the program package CrysAlisPro 1.171.38.41. The structure was solved by Direct Methods using the SIR97 program and refined on F_o^2 by full-matrix least-squares procedures, using the SHELXL-2014/7 program in the WinGX suite v.2014.1.[46] All non-hydrogen atoms were refined with anisotropic atomic displacements. The carbon-bound H atoms were placed in calculated positions and refined isotropically using a riding model with C-H ranging from 0.97 to 0.98 Å and U_{iso}(H) set to $1.2U_{eq}$(C). The O-bound H atoms of the water molecules were found in the difference Fourier and then refined riding with O-H distances ranging from 0.85 to 0.87 Å and U_{iso}(H) set to $1.5U_{eq}$(O). Several disordered water molecules could not be sensibly modelled in terms of atomic sites, and were treated using the PLATON SQUEEZE procedure. Their contribution to the diffraction pattern was removed and modified F_o^2 written to a new HKL file. The number of electrons corresponding to the solvent molecules were included in the formula, formula weight, calculated density, μ and F(000). SQUEEZE calculated a Solvent Accessible Volume of 3170 Å^3 and 933 electrons per unit cell. From the inspection of the Fourier map, these were attributed to 80 water molecules per cell (10 water molecules in the asymmetric unit). The weighting schemes used in the last cycle of refinement was $w = 1/[\sigma^2 F_o^2 + (0.5167P)^2]$ where $P = (F_o^2 + 2F_c^2)/3$.

For CBT-2 intensity data and cell parameters were recorded at 200(2) K on a Bruker D8 Venture PhotonII diffractometer equipped with a CCD area detector, using the MoKa radiation. The raw frame data were processed using the programs SAINT and SADABS. The structure was solved by Direct Methods using the SIR97 program and refined on F_o^2 by full-matrix least-squares procedures, using the SHELXL-2014/7 program in the WinGX suite v.2014.1.[46] All non-hydrogen atoms were refined with anisotropic atomic displacements, with the exception of some oxygen atoms of the lattice water molecules. The carbon-bound H atoms were placed in calculated positions and refined isotropically using a riding model with C-H ranging from 0.99 to 1.00 Å and $U_{iso}(H)$ set to 1.2 $U_{eq}(C)$. The O-bound H atoms of the water molecules could neither be found in the difference Fourier nor calculated from the H-bonding network, with the exception of the water molecule O5W. The H atoms were refined isotropically using a riding model, with O-H distances of 0.87 Å and $U_{iso}(H)$ set to $1.5U_{eq}(O)$. Several disordered water molecules could not be sensibly modelled in terms of atomic sites, and were treated using the PLATON SQUEEZE procedure. Their contribution to the diffraction pattern was removed and modified F_o^2 written to a new HKL file. The number of electrons corresponding to the solvent molecules were included in the formula, formula weight, calculated density, μ and F(000). SQUEEZE calculated a Solvent Accessible Volume of 1776 Å^3 and 645 electrons per unit cell. From the inspection of the Fourier map, these were attributed to 64 water molecules per cell (8 water molecules in the asymmetric unit). The weighting schemes used in the last cycle of refinement was $w = 1/[\sigma^2 F_o^2 + (0.4538P)^2]$ where $P = (F_o^2 + 2F_c^2)/3$.

Parameter	CBT-1	CBT-2
Empirical formula	$C_{48}H_{95}N_{32}O_{38}Cl_2I$	$C_{48}H_{152}N_{32}O_{64}Cl_8$
Formula weight	1926.35	2485.61
Crystal system	Orthorhombic	Monoclinic
Space Group	$Pccn$	$C2/m$
a/Å	26.1231(3)	31.9007(8
b/Å	26.0971(2)	10.6675(2)
c/Å	12.93180(10)	18.0682(4)
β/deg	-	113.949(1)
V/Å^3	-	5619.3(2)
Z	4	2
T/K	100(2)	200(2)
ρ/cm^3	1.449	1.469
μ/mm^{-1}	0.355	0.312
F(000)	3980	2624
total reflections	166193	62581
unique reflections (R_{int})	14717 (0.0375)	9679 (0.0419)
observed reflections $[F_o/ > 4/\sigma(F_o)]$	1.006	0.998
GOF on F^{2a}	1.006	0.998
R indices $[F_o > 4\sigma(F_o)]^b R_1, wR_2$	0.1367, 0.4892	0.1422, 0.4352
largest diff. peak and hole (eÅ^3)	4.359, -2.324	4.699, -1.387

[a]Goodness-of-fit $S = [\Sigma(F_o^2 - Fc2)2/(n - p)]1/2$, where n is the number of reflections and p the number of parameters. [b]$R1 = \Sigma||F_o| - |F_c||/\Sigma|F_o|, wR_2 = [\Sigma[w(F_o^2 - F_c^2)^2]/\Sigma[w(F_o^2)^2]]^{1/2}$.

Table 3.1: Crystallographic data for tubular CB[8] assemblies

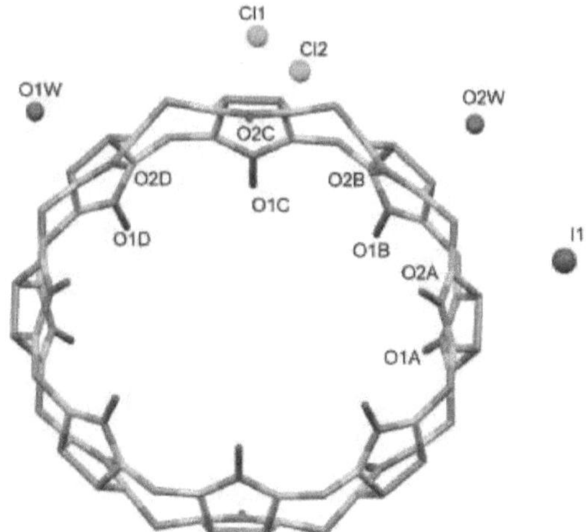

Figure 3.21: Molecular structure of the asymmetric unit of CBT-1 with the whole macrocycle generated by symmetry. The oxygen atoms of the water molecules and the chlorine/iodine anions are represented as red spheres of arbitrary radius. The symmetry code to generate the entire CB[8] is 1-x, 1-y, 1-z. H atoms have been omitted for clarity.

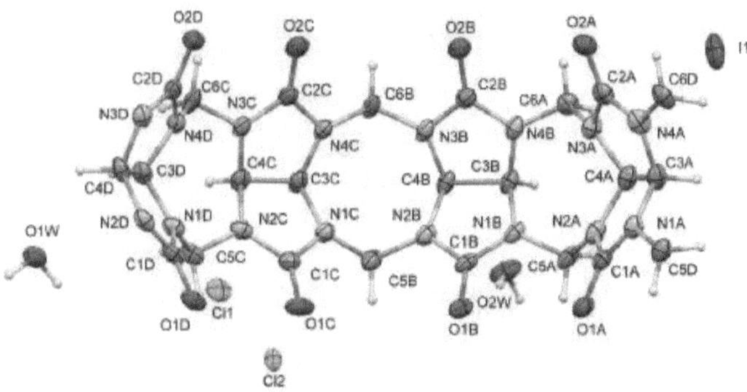

Figure 3.22: ORTEP view of the asymmetric unit of CBT-1 with ellipsoids drawn at the 20% level. The water molecules have occupancy factor 0.5, as well as Cl1 and Cl2.

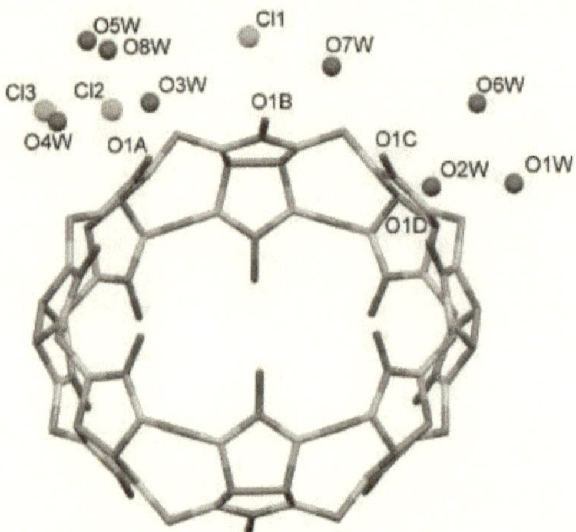

Figure 3.23: Molecular structure of the asymmetric unit of CBT-2 with the whole macrocycle generated by symmetry. The oxygen atoms of the water molecules and the chlorine anions are represented as red and green spheres of arbitrary radius. The symmetry codes to generate the entire CB[8] are 1-x, y, 2-z; 1-x, 2-y, 2-z; x, 2-y, z. H atoms have been omitted for clarity.

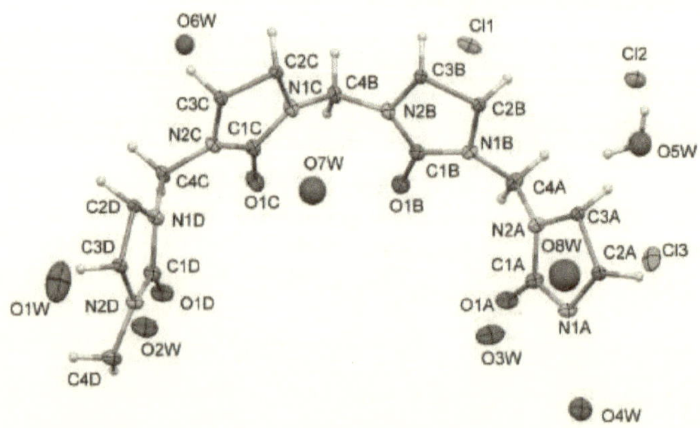

Figure 3.24: ORTEP view of the asymmetric unit of CBT-1 with ellipsoids drawn at the 20% level. The water molecules have occupancy factor 0.5, as well as Cl1 and Cl2.

X-ray crystallographic section for CB[8] - $ZnCl_4^{2-}$ assembly

The crystal structures of the compound was determined by X-ray diffraction methods. Crystal data and experimental details for data collection and structure refinement are reported in Table 3.2. Intensity data and cell parameters were recorded at 200(2) K on a Bruker D8 Venture Photon II diffractometer equipped with a CCD area detector, using the CuKa radiation ($\lambda = 1.54178$). The raw frame data were processed using the programs SAINT and SADABS. The structure was solved by Direct Methods using the SIR97 program[47] and refined on F_o^2 by full-matrix least-squares procedures, using the SHELXL-2014/7 program[46] in the WinGX suite v.2014.1.[48] All non-hydrogen atoms were refined with anisotropic atomic displacements. The carbon-bound H atoms were placed in calculated positions and refined isotropically using a riding model with C-H ranging from 0.99 to 1.00 Å and $U_{iso}(H)$ set to $1.2U_{eq}$. Some of the O-bound H atoms of the water molecules could neither be found in the difference Fourier nor calculated from the H-bonding network. Those which could be calculated from the H-bond network were refined isotropically using a riding model, with O-H distances of 0.87 Å and $U_{iso}(H)$ set to $1.5U_{eq}(O)$. Several disordered water molecules could not be sensibly modelled in terms of atomic sites, and were treated using the PLATON SQUEEZE procedure.[49] Their contribution to the diffraction pattern was removed and modified F_o^2 written to a new HKL file. The number of electrons corresponding to the solvent molecules were included in the formula, formula weight, calculated density, μ and F(000). SQUEEZE calculated a Solvent Accessible Volume of 1415 Å^3 and 412 electrons per unit cell. From the inspection of the Fourier map, these were attributed to 20 water molecules per cell (10 water molecules in the asymmetric unit). The weighting scheme used in the last cycle of refinement was $w = 1/[\sigma^2 F_o^2 + (0.2986P)^2]$ where $P = (F_o^2 + 2F_c^2)/3$.

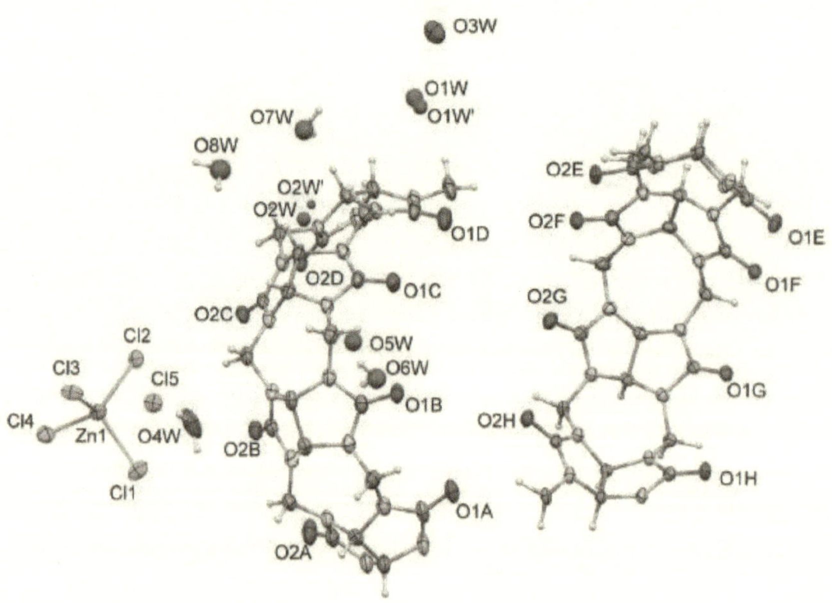

Figure 3.25: ORTEP view of the asymmetric unit of $ZnCl_4^{2-}$-CB[8] with partial labelling scheme and ellipsoids drawn at the 20% level

Crystallographic Data for CB[8] - $ZnCl_4^{2-}$ assembly

Parameter	Value	
Empirical formula	$C_{48}H_{83}N_{32}O_3ZnCl_5$	
Formula weight	1863.08	
Crystal system	Triclinic	
Space Group	$P-1$	
a/Å	17-1817(10)	
b/Å	17.7959(11)	
c/Å	17.8354(11)	
α/deg	71.424(3)	
β/deg	82.393(4)	
γ/deg	64.781(3) V/Å^3	4676.5(5)
Z	2	
T/K	200(2)	
ρ/cm^{-1}	1.323	
μ/mm^{-1}	2.443	
F(000)	1932	
total reflections	86001	
unique reflections (R_{int})	16017 (0.0942)	
observed reflections $[F_o/ > 4/\Sigma(F_o)]$	10515	
GOF on F^{2a}	1.0212	
R indices $[F_o > 4\sigma(F_o)]^b R_1, wR_2$	0.1178, 0.3461	
largest diff. peak and hole eÅ^{2a}	1.898, -0.440	

[a]Goodness-of-fit $S = [\Sigma(F_o^2 - Fc2)2/(n-p)]1/2$, where n is the number of reflections and p the number of parameters. [b]$R1 = \Sigma||F_o| - |F_c||/\Sigma|F_o|, wR_2 = [\Sigma[w(F_o^2 - F_c^2)^2]/\Sigma[w(F_o^2)^2]]^{1/2}$.

Table 3.2: Crystallographic Data for $ZnCl_4^{2-}$-CB[8] assembly

Chapter 4

Different encapsulation modes of guests within CB[8]

4.1   Introduction

CB[8] has the ability to form inclusion complexes of different stoichiometry and stability based on the nature of the guest molecules and their concentration in solution. In this chapter several complexes of CB[8] and guest molecules were studied, and X-ray diffraction studies of a series of complexes ranging from 1:1, 2:1, and 2:2 with respect to the host were carried out.

The photophysical properties of organic dye molecules are very sensitive to their environment. As a result, fluorescent dyes have found countless applications as molecular probes. As briefly discussed in Chapter 2, complexation of fluorescent guest molecules by macrocyclic hosts can drastically change their microenvironments subsequently affecting their fluorescence. Particularly in the case of CB[8], more than one dye molecule can be included within the hydrophobic cavity in aqueous solution, often leading to some interesting spectroscopic changes. Several dye-CB[8] complexes were studied in order to select a system suitable for use as a self diagnostic probe in damage reporting polymers, and the results are reported in this chapter.

4.2 Cyanine 3 - CB[8] 1:1 complex

Cyanine dyes are a class of dyes in which two heterocyclic groups such as imidazole, pyridine and indole are interconnected by a polymethine chain.[1] A major advantage of these dyes is the easy tuning or manipulation of spectral properties during synthesis by changing the polymethine chain length and insertion of functional groups on the heterocycles (Figure 4.1). In the early 1990's, cyanine dyes became commercially available in the succinimidyl ester form, with increased water solubility and reduced fluorescence quenching by dye-dye interactions. This ease of availability kickstarted their widespread use as labelling agents for nucleic acids.

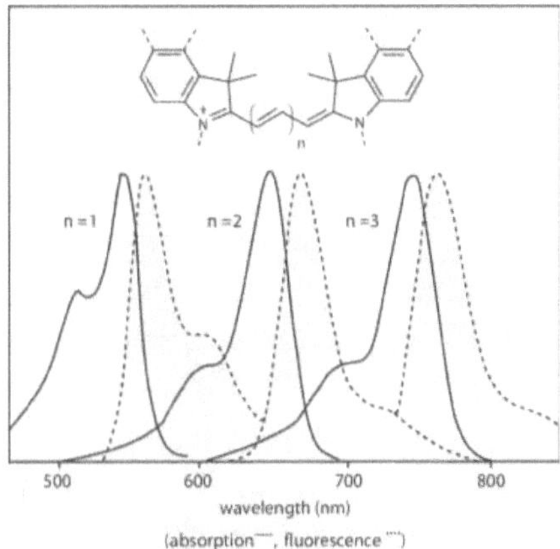

Figure 4.1: Spectroscopic properties of cyanine dyes can be changed by varying the polymethine chain length (Adapted from reference [2]).

Currently there is huge interest in using organic dyes such as cyanines, rhodamines and fluorescenes as probes for a variety of biomolecules for bioimaging and their study by different fluorescence detection techniques.[3, 4] However one of the drawbacks is that these molecules often undergo rapid photobleaching and present low quantum yields.[5, 6] Supramolecular host-guest complexation by molecules such as cyclodextrins and cucurbiturils has been shown to cause the alteration of the microenvironment of dyes leading to changes in their spectral properties.[7–9] Additionally, by tuning the structure of the cyanine dyes it is possible to obtain absorption and emission in the near IR range. This offers better depth penetration of light. This could be very advantageous to the design of supramolecular probes for damage detection in materials, and could allow us to detect microdamage deeper below the surface than possible with the current system. This was our main interest in studying the complexation of cyanine dyes with CB[8].

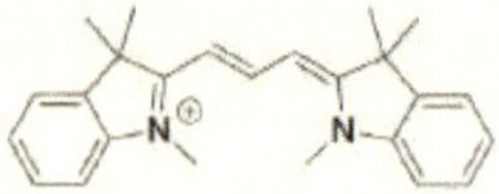

Figure 4.2: Structure of dye Cyanine-3 (Cy3).

We investigated the interaction of a model indocyanine dye cyanine-3 (Cy3) with CB[8]. The structure is shown in Figure 4.2. The dye was synthesised following a two step procedure reported in the literature.[10]. l-methyl-2,3,3,-trimethyl-3H-indolium iodide was synthesised by refluxing 2,3,3-trimethyl-3H-indole with iodomethane at 130 °C for 6 h. Subsequently the red product was reacted with triethylorthoformate in pyridine to give the dye Cy3 which was purified by recrystallisation in ethanol. The synthetic scheme of the final step is shown in Scheme 4.1.

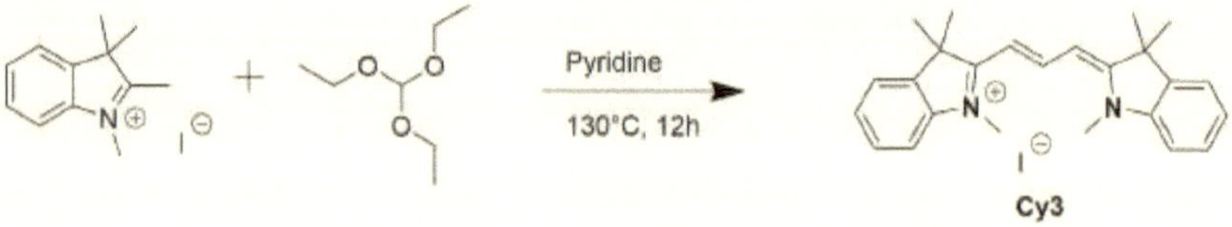

Scheme 4.1: Synthetic scheme for dye Cy3.

4.2.1 Spectroscopic changes upon encapsulation

UV-Vis and fluorescence spectra

The behaviour of the dye upon addition of CB[8] in aqueous solution was studied at a concentration of 5 µM to avoid dye self aggregation. Cy3 in aqueous solution shows two UV absorption maxima at 510 nm and 541 nm. As shown in Figure 4.3, upon addition of CB[8] a red shift was observed in the λ_{max} along with an isosbestic point at 545 nm and a slight decrease in absorption coefficient as shown in Figure 4.3. This pronounced spectral shift suggests a change in the microenvironment of the dye within the hydrophobic cavity of the CB[8]. The absorption at 541 nm steadily decreased with every addition of CB[8], and remained steady after addition of one equivalent of CB[8].

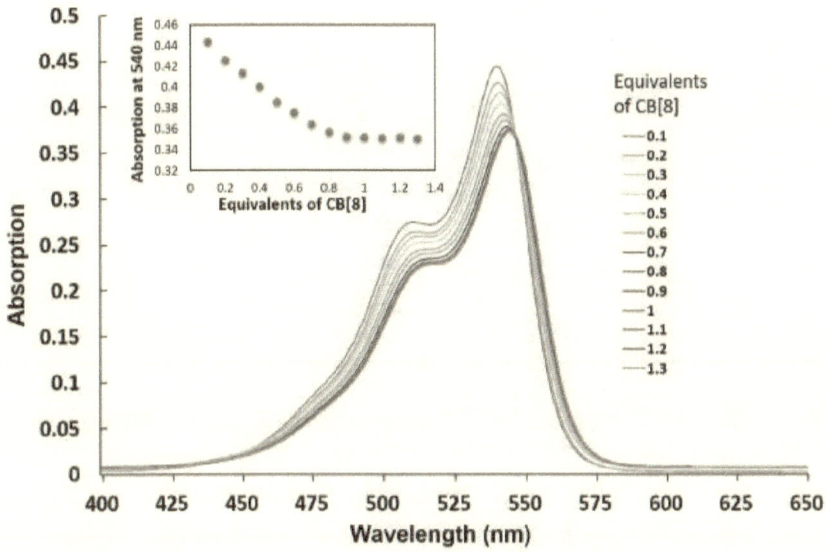

Figure 4.3: UV Visible spectrum of Cy3 in aqueous solution upon addition of CB[8] (5 μM). The inset shows change in absorption at 541 nm against equivalents of CB[8].

Adding more than one equivalent of CB[8] does not cause significant changes in the spectrum as shown in the inset in Figure 4.3.

Furthermore, the effect of the fluorescence emission of Cy3 in the presence of CB[8] was investigated as shown in Figure 4.4. Upon successive additions of CB[8], the fluorescence intensity decreases and a spectral shift was observed.

Cyanine dyes have a tendency to lose their fluorescence upon torsional motion.[11, 12] This phenomenon can be clearly observed in the fluorescence spectrum of Cy3 in solvents with different viscosity. In fluid solutions, these

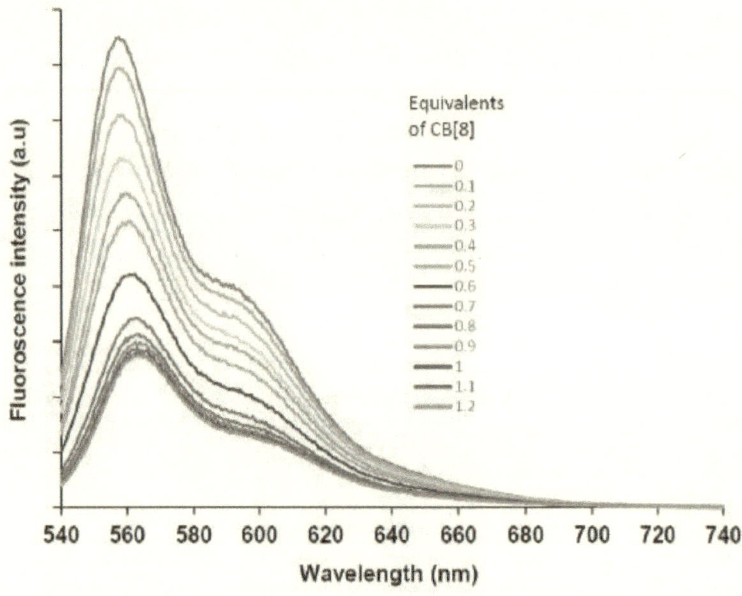

Figure 4.4: Fluorescence spectrum of Cy3 in water (5 µM) upon addition of CB[8], $\lambda_{ex} = 500$ nm.

dyes typically exhibit very low fluorescence intensity, while viscous solvents such as glycerol promote substantial enhancement in fluorescence. In a 95% methanol:water solution, the fluorescence intensity of the Cy3 dye was much lower than in water. On the contrary, in a 95% ethylene glycol:water solution the fluorescence intensity is drastically increased. The harder the torsional motion and isomerisation of the dye, and higher the fluorescence intensity (Figure 4.5).

The quenching of fluorescence of the guest molecule upon encapsulation with the CB[8] is unprecedented for single guest encapsulation. As discussed earlier, the behaviour of cyanine dyes is affected by the ability of the excited state to isomerise into non emissive geometries. One reason for quenching upon binding could be as follows. Since one indolenine moiety is bound to the CB[8], it is possible that binding within CB[8] partially separates the molecule from the bulk solvent by isolating it within the cavity and subsequent release of

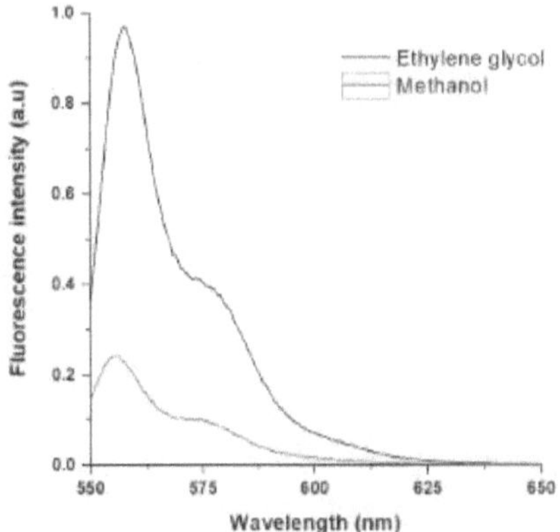

Figure 4.5: Fluorescence spectrum of Cy3 in solvents with different viscosity (5 µM), λ_{ex} =500 nm

high energy water molecules. This could in turn enable increased rotor motion and subsequently cause the excited state to deactivate increasingly through a non-radiative channel, leading to a loss in florescence intensity.

^{1}H NMR titration

The binding interaction between Cy3 and CB[8] was also monitored using ^{1}H-NMR spectroscopic data recorded in D_2O. Figure 4.6 shows the difference between the spectrum of Cy3 and the spectrum of Cy3 and CB[8] (1:1) at 0.1 mM. A sizable upfield shift was observed for the aromatic protons ($H_{1,2,3,4}$), while slight upfield shifts were also seen for the signals of the polymethine chain ($H_{7,8}$). The largest perturbation was observed for the protons in the aromatic region. Since upon binding the positive charge should be localised on the heterocycle closer to the portals, it was notable that the spectrum still showed symmetric signals for the molecule Cy3. Overall, this suggests a fast exchange of the complex where the most probable scenario is a shuttling of the CB[8] between the two aromatic regions of the molecule. Additionally, when the ^{1}H NMR spectrum was recorded at 1:1 equivalents of CB[8], the signals of Cy3 were significantly broadened and were barely visible. This absence of sharp and dispersed signals also suggests the formation of a supramolecular polymer in solution over time as observed by Scherman and coworkers in other CB[8] complexes.[13]

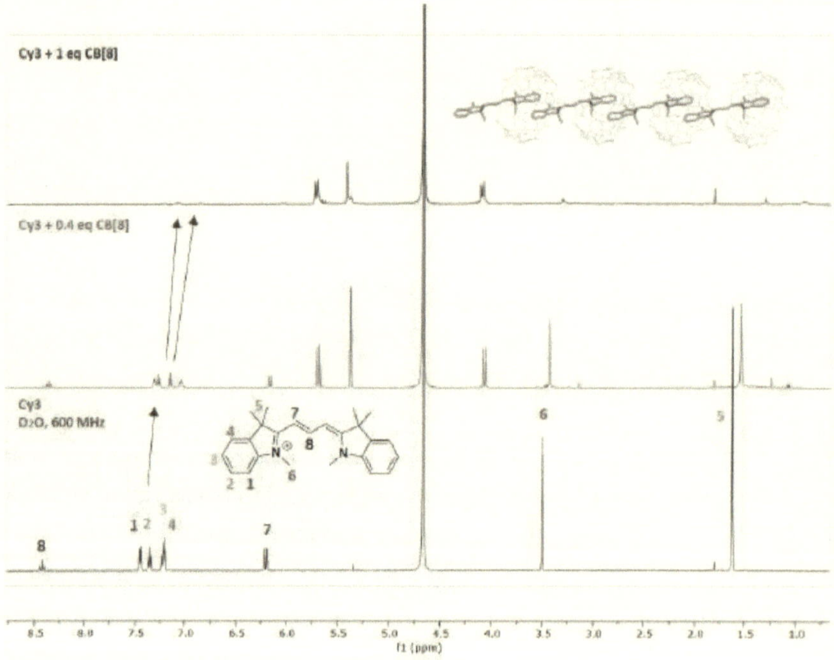

Figure 4.6: ^{1}H NMR (600 MHz, D$_2$O): Addition of increasing amounts of CB[8] to free Cy3 (red trace) at 0.1 mM of Cy3.

Isothermal titration calorimetry

An isothermal titration calorimetry (ITC) experiment was then performed to investigate the binding strength, stoichiometry and thermodynamic parameters for the host-guest complexation of Cy3 by CB[8]. Figure 4.7 shows a typical ITC titration experiment with Cy3 and CB[8]. A 27 injection experiment was carried out at $T = 25\,°C$ where the cell contains a 0.1 µM aqueous solution of CB[8], and Cy3 is in the syringe at a concentration of 1.5 µM.

The observed exothermic binding curve has a sharp inflection point at a molar ratio of one indicating the formation of a 1:1 assembly in aqueous solu-

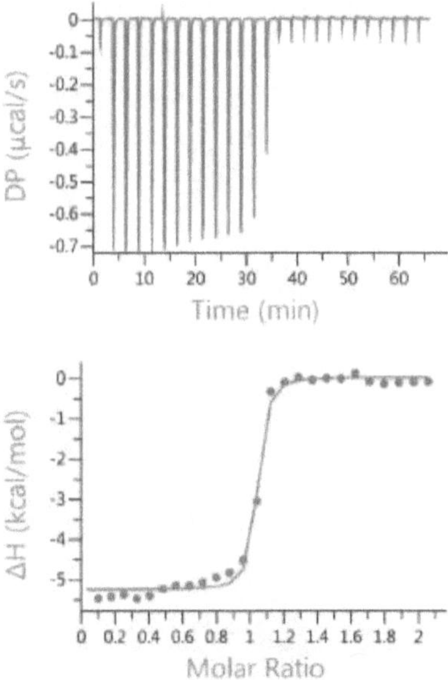

Figure 4.7: Isothermal calorimetric data obtained in the titration of a solution CB[8] by adding increments of the Cy3 monomer [CB[8]]=0.1 mM, [Cy]=1.5 mM

tion. The data fitting of the titration to a 1:1 binding model gave an association constant $K_a = 1.21 \times 10^7$ M^{-1}. The parameters calculated were ΔH = -20.50 kJ mol^{-1}, ΔG = -40.50 kJ mol^{-1} and TΔS = 20.04 kJ mol^{-1} making this an example of both enthalpy and entropy driven binding in CB[8]. The enthalpic contribution arises from the displacement of high energy water molecules from the cavity by the guest, which is known to be one of the main driving forces of CB[8] complexation.[14]

4.2.2 Supramolecular polymer in solid state

While most complexes of CB[8] show a much higher solubility in water than the CB[8] itself, it was surprising to see that during solution studies of binding of the dye Cy3 to CB[8], upon the addition of one equivalent of CB[8] over the course of a few hours the complex started to precipitate as bright purple and distinctly needle shaped crystals. In order to understand the arrangement of

Cy3 in the cavity of CB[8], crystals suitable for X-ray diffraction were grown in water. Single crystal X-ray diffraction confirmed the 1:1 binding observed upon spectroscopic investigation in solution where the dye forms a complex with CB[8] with one indolinene 'head' of the dye positioned inside the hydrophobic cavity as shown in Figure 4.8. However in the solid state, the tail of the indolenine also interacts with an adjacent cucurbituril, leading to the formation of a linear supramolecular polymer as shown in Figure 4.9.

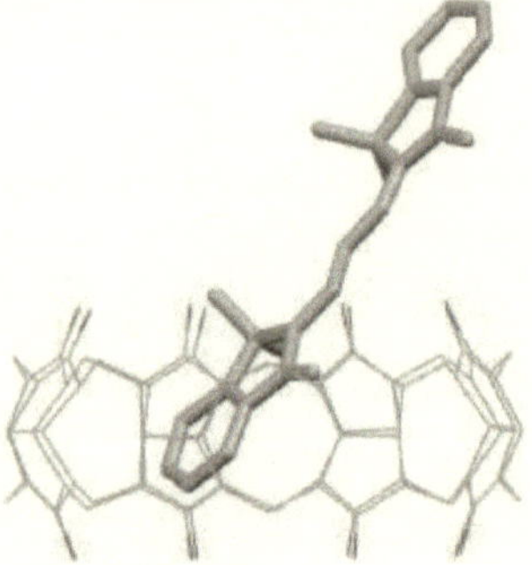

Figure 4.8: One CB[8] unit encapsulating Cy molecules in a "head to tail" manner.

The structure is currently under refinement but a preliminary analysis shows that each Cy3 molecule is positioned in the CB[8] cavity in a way that one heterocyclic region is positioned deep within the hydrophobic cavity. To the same CB[8], the tail of a neighbouring Cy3 is also coordinated through interactions with the carbonyl portals as well as weak π-π interaction with the first molecule. In this way a linear chain like assembly is formed, with each Cy3 encapsulated by two CB[8] molecules as shown in Figure 4.9. It is also interesting to note the tilted orientation of the main guest molecule within the cavity.

Each chain of the supramolecular polymer is arranged parallel to each other and offset by half the width occupied by one CB[8]. Similar to the first columnar assembly reported in Chapter 3 (CBT-1), the linear chains are made of bumps and hollows. Therefore each neighbouring chain is positioned such that the CB[8] occupies the hollows of the first chain as shown in Figure 4.10.

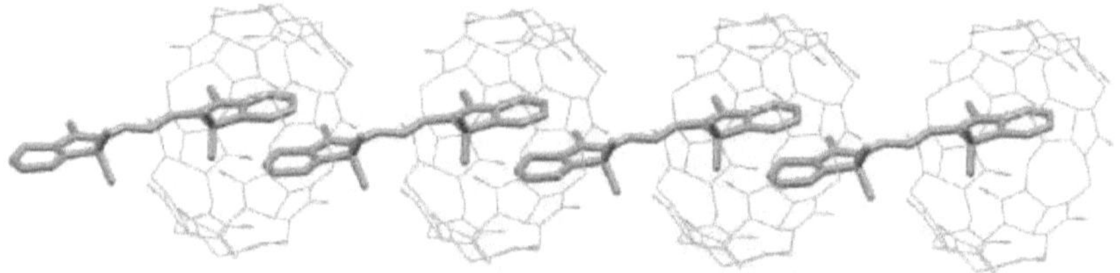

Figure 4.9: Linear polymeric structure formed by CB[8] and Cy3 along the c axis.

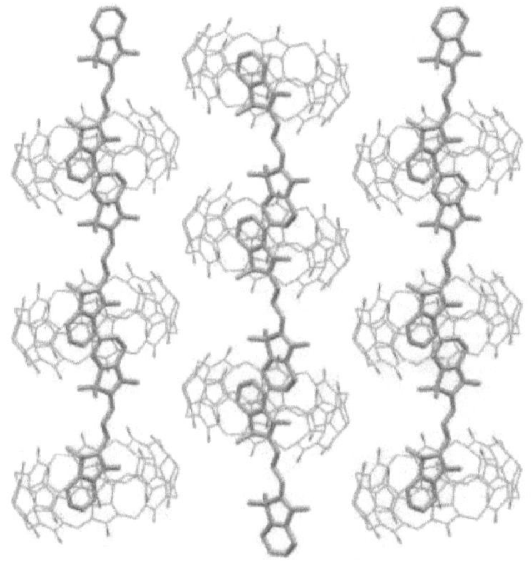

Figure 4.10: Parallel orientation of linear polymeric chains viewed along the b axis.

It is also interesting to note the structural similarity of the columnar arrangement of the Cy3-CB8 complex with the nanotubular framework CBT-1. The angle of tilt of each CB[8] is identical. The only difference is a slight increase in the distances of consecutive CB[8] molecules in order to accommodate the guests molecules. This head to tail arrangement of Cy3 is different to that observed by crystals of Cy3 alone grown in aqueous solution, shown in Figure 4.11. Overall the stoichiometry of the complex is the same (1:1) both in solution and in the solid state.

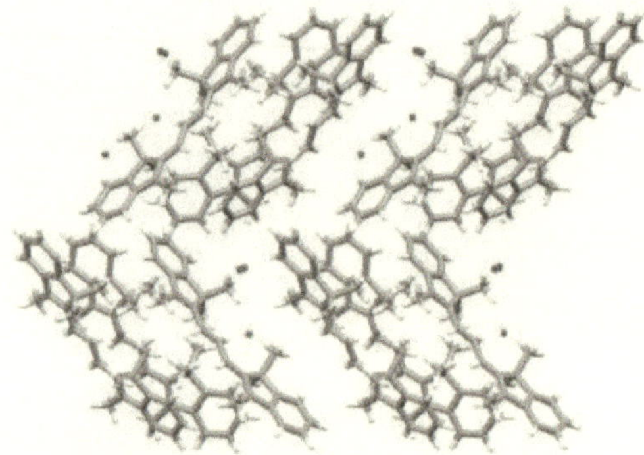

Figure 4.11: Cy3 packing in solid state without CB[8]

4.3 4,4' Azopyridine-CB[8] 2:1 complex

4,4'-Azopyridine (Figure 4.12) is part of a series of N-donor heterocycles containing two pyridyl units that has been extensively employed as a building block in coordination compounds and co-crystals.

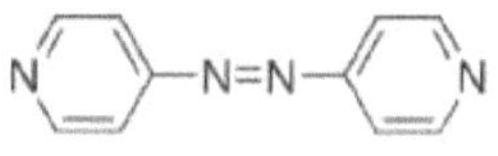

Figure 4.12: Structure of 4,4'-azopyridine

Compared to azobenzene, azoheteroaromatic systems have been less frequently examined, even though some of these molecules share important properties with azobenzene derivatives. For example, symmetrical azopyridines (2,2'-azopyridine, 3,3'-azopyridine and 4,4'-azopyridine) have cis–trans activation energies on the order of the azobenzene analogue (i.e., 21.0, 22.6 and 21.4 kcal mol^{-1} in solution, respectively).[15] The potential for hydrogen bonding to nitrogen lone pairs in these molecules gives rise to the prospective study of azoheteroaromatic derivatives as hydrogen bond arrays. In this regard, the use of heteroaromatic moieties (such as pyridyl, pyrimidyl, and naphthyridyl) and a nitrogen atom from the azo group have been explored as hydrogen bond acceptor arrays in molecular recognition.[16, 17]

^{1}H NMR Titration

The protonation or alkylation of the nitrogen groups on the azopyridine should cause the molecule to behave as a first guest for CB[8], similar to methyl viologen. Upon NMR titration with CB[8] in D_2O, the chemical shift pertubation of the azopyridine confirmed complex formation. Two new peaks corresponding to the aromatic protons on the encapsulated CB[8] were observed, shifted upfield by $\Delta\delta = 1.3$ ppm. The two species were found to be in slow exchange, and upon addition of 0.5 equivalent of CB[8] the signals of the free azopyridine were barely visible. In addition there was a clear splitting of the protons of the host CB[8]. The mode of association of AzPy with CB[8] depends highly on the protonation state of the pyridyl nitrogens. A 2:1 binding points to a monoprotonation of the AzPy in solution.

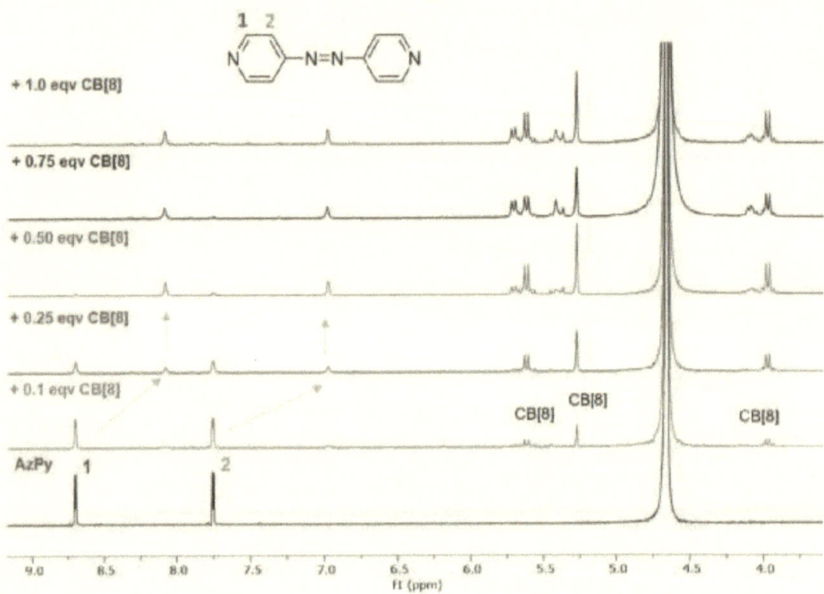

Figure 4.13: ^{1}H NMR titration of AzPy with CB[8] in D$_2$O, 600 MHz at a concentration c[AzPy] = 0.2 mM. The concentration of CB[8] is varied from 0 to 0.15 mM

An additional interest in the interaction between CB[8] and azopyridine was in the ability of the protonated nitrogen moiety to participate in additional noncovalent interactions along with the hydrophobic interactions in the cavity. In order to understand whether these interactions could play a role in the solid state packing of the complex we attempted to grow single crystals of the complex in water. When a solution of CB[8] and azopyridine was subjected to hydrothermal treatment in a close sintered glass vial for 48 h and then left at room temperature, red crystals were observed in the solution. Single crystal X-ray diffraction analysis revealed that the azopyridines were monoprotonated, leading to the formation of dimers within the CB[8] cavity as shown in Figure 4.14.

The mode of association of AzPy with CB[8] is dependant on the protonation state of the pyridyl nitrogens. In the diprotonated state, azopyridine is

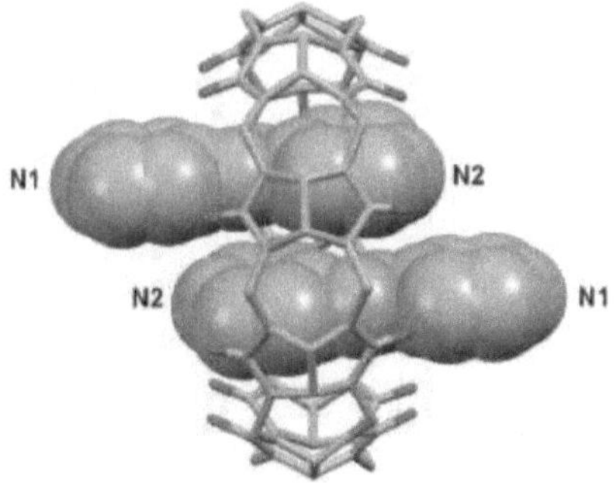

Figure 4.14: Perspective view of the supramolecular complex formed by CB[8] and two molecules of 4,4'-azopyridine (in space filling mode). H atoms have been omitted for clarity.

expected to form a 1:1 complex with CB[8] while in the monoprotonated state the dye is able to form homodimers within the cavity. Even after repeated washing of the CB[8] to remove residual acid from the synthesis and purification process, the solution of CB[8] always has a pH below 5. Under these conditions it is reasonable that the molecule undergoes protonation at at least one pyridyl moiety. This proton is hydrogen bonded to the carbonyl oxygens of the adjacent CB[8] molecule, and is the reason for the formation of the 2:1 polymeric structure observed in the solid state as shown in Figure 4.15.

In the structure, the two molecules of azopyridine are sandwiched inside its cavity mainly through $C-H_{(AP)}\cdots O_{(CB)}$ and dispersion interactions. Additionally, the protonated nitrogen on the pyridine protruding out of the cavity is hydrogen bonded to two oxygen atoms O1A and O1B of the neighbouring CB[8]. The combined effect of these interactions results in the formation of supramolecular chains which extend along the c-axis direction of the unit cell (Figure 4.16).

Description of the crystal structure

The crystal structure of compound AzPy-CB[8] was determined via X-ray diffraction data on single crystals obtained by slow evaporation of a 0.1 mM

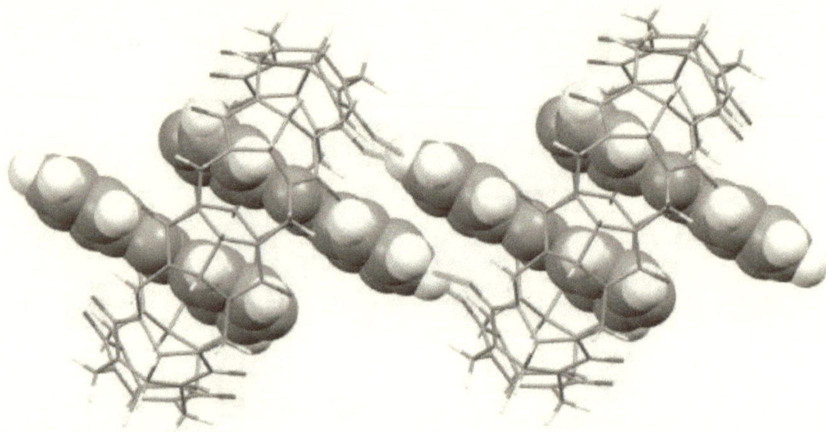

Figure 4.15: H-bonds between AzPy and adjacent CB[8] carbonyl oxygens are shown as blue dashed lines

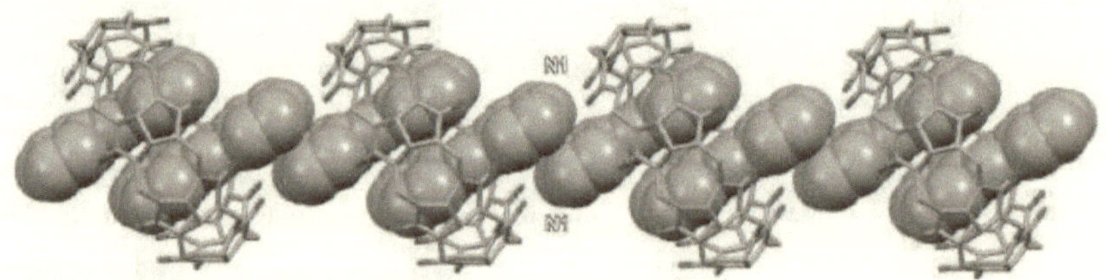

Figure 4.16: One of the supramolecular chains of CB[8] and 4,4'-azopyridine molecules (in space-filling mode) formed along the *c*-axis direction of the unit cell. H atoms have been omitted for clarity

solution of AzPy and CB[8].

The asymmetric unit (see Figure 4.17) consists of half a CB[8] macrocycle, four water molecules (eight molecules each of them displaying an occupancy factor of 0.5), one and a half nitrate anion and one 4,4'-azopyridine which is fully protonated on the N atom N1 and "half" protonated on the N atom N2, to balance the half negative charged carried by the half nitrate counterion N6O4O5O6.

The presence of a proton on N2 is also inferred by the very short N2···O1, N2···O3 and N2'···O1 distances of 2.611(9), 2.952(9), and 2.892(9) Å respec-

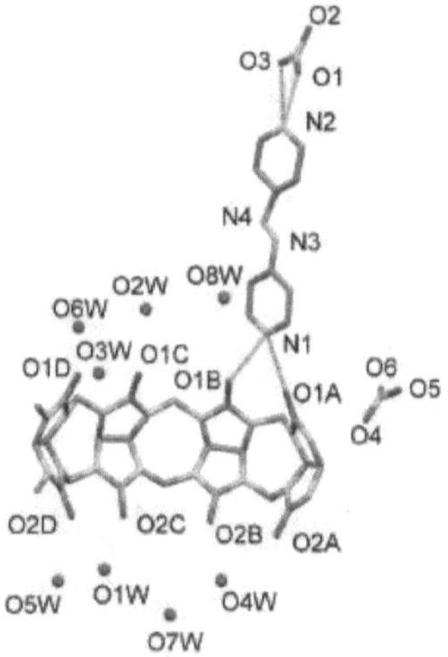

Figure 4.17: Molecular structure of the asymmetric unit of AzPy-CB[8] with partial labelling scheme. Hydrogen atoms and the second orientation of the pyridine ring have been omitted for clarity. The symmetry code to generate the whole molecule is 1-x, 1-y, -z. H bonds are shown as cyan dotted lines and the oxygen atoms of the water molecules as red spheres of arbitrary radius.

tively (N2' is the nitrogen atom belonging to the second orientation of the pyridine ring C6-C10/N2, which is disordered over two equivalent positions). This interaction is also marked in Figure 4.17, along with the H bonds formed by the protonated N1-H1N fragment with the oxygen atoms O1A and O1B of CB[8] [N1$-$H1N$\cdots$O1A: 2.783(.004) Å and 133.4(2)°; N1$-$H1N$\cdots$O1B: 2.900(6) Å and 128.5(3)°]. Weak π-π contacts between symmetry related C1-C5/N1 pyridine rings [Cg$\cdots$Cg^i 3.672(2) Å ; i = 1-x, 1-y, 1-z] are also present.

In fact, the 4,4' dimethyl azopyridine diiodide (AzPy-2Me) counterpart was also synthesised. However when crystals were grown from a solution of the AzPy-2Me and CB[8], it was observed that there was no inclusion of the guest and a nanotubular framework CBT-1 was observed similar to the case of methyl viologen described in Chapter 3. The hydrogen bonding of the protonated pyridine to the neighbouring CB[8] is therefore essential to prevent the slipping out of the guest during the crystallisation process.

Isothermal Titration Calorimetry of AzPy-CB[8] complex

An ITC experiment was then performed to investigate the binding of AzPy by CB[8]. The experiments were carried out in acetate buffer at a pH of 4.75, close to the pKa of azopyridine to ensure that the monoprotonated species would dominate. Figure 4.18 shows the ITC titration experiment with AzPy and CB[8] in 50 mM sodium acetate buffer at pH = 4.75. Experiments were carried out at $T = 25\,°C$, where the cell contained a 0.1 µM solution of CB[8], and the syringe contained AzPy at $[c] = 1.5$ µM.

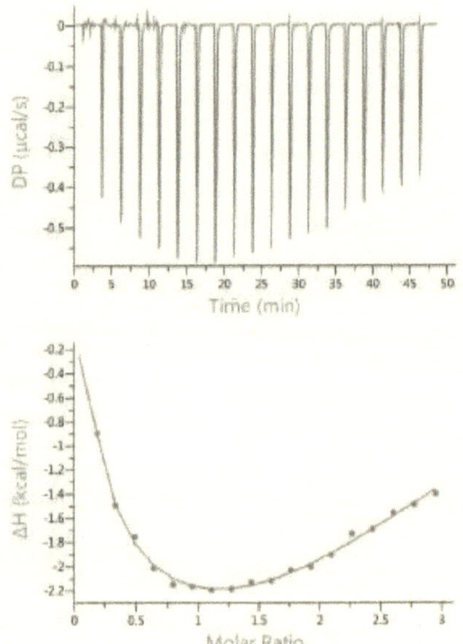

Figure 4.18: Isothermal calorimetric data obtained in the titration of CB[8] in 50 mM acetate buffer pH = 4.75 by adding increments of the AzPy monomer [CB[8]]=0.1 mM, [AzPy]=1.5 mM

Fitting the data using a sequential binding model where N = 2 sites gave association constant $K_{a1} = 0.43 \times 10^4$ M^{-1} and $K_{a2} = 3.13 \times 10^5$ M^{-1}, and the binding was enthalpy driven. Since the second association of the AzPy is higher than the first the complex shows an example of coorporative binding, which can be explained as follows. As reported by Scherman and co-workers, the release of high energy water - geometrically confined water molecules within the cavity - can provide an enthalpic driving force that is regarded as a nonclassical hydrophobic effect.[14] After the encapsulation of the first guest, the space within the cavity is further reduced. This geometric confinement coupled with the hydrophobic nature of the guest within the molecule increase the energy of the residual waters, making the second binding more favourable.[18]

The solution state experiments and solid state behaviour are coincident and show a 2:1 homodimeric ternary complex with CB[8].

4.4 Encapsulation of alkylated viologen with CB[8] - 1:2 binding in solid state.

While azopyridines and bipyridines form stable complexes with CB[8] depending on the protonation state of the pyridyl nitrogen, alkylated bipyridines also called viologens, with their permanent charge show exceptional binding with CB[8] independent of pH. Viologen dication derivatives have therefore been well studied as a class of guest molecules for CB[7] and CB[8] and their strong association with these hosts can be attributed to the contributions of the favourable ion–dipole interaction between the positive charges of the guest and portal oxygen atoms of CB[8] as well as the hydrophobic effects of the cavity.[19] In the case of CB[8], even though the cavity is large enough to accommodate two MV^{2+} molecules due to charge repulsion they form stable 1:1 complexes. For this reason the most commonly used as 'first guests' for heteroternary complexation.[20] However, the binding stoichiometry of a host–guest complex could be effectively controlled by the redox chemistry of the guest. Kim *et al.* showed that the 1:1 inclusion complex of methylviologen dication (MV^{2+}) in CB[8] converted completely and reversibly to a 2:1 inclusion complex of the cation radical (MV^{2+}) in CB[8] upon the reduction of the guest leading to the formation of high affinity CB[8] homoternary complex.

Around the same time, they also reported for the first time an unusual U-shaped conformation of alkyl chains encapsulated within the cavity CB[8].[21, 22] The discovery of this back-folding of the alkyl chains in a host later led to the design and synthesis of a molecular machine based on the back-folding of the alkyl chains and host-stabilised charge transfer complex formation in CB[8].[23]

In the previous chapter, we describe the use of a dicationic viologen derivative as a chaperone molecule in the synthesis of tubular frameworks of CB[8]. As a control experiment, we attempted the crystallisation of CB[8] with a longer chain viologen derivative terminated with a double bond, MVC_{11}. Our presumption was that the longer chain would dissuade the close packing of the CB[8] as observed with methyl viologen, and not provide the chaperone effect observed earlier. As expected, the tubular packing was not observed, however when the crystals obtained by this experiment were examined by single crystal X- ray diffraction, an unusual complex was observed comprising of the ternary complex of the host and guest, but contrary to the usual observation there were two molecules of CB[8] binding each viologen as shown in Figure 4.19. One CB[8] encapsulates the aromatic head of the molecule as expected, while a second molecule of CB[8] binds the tail alkyl chain.[21]

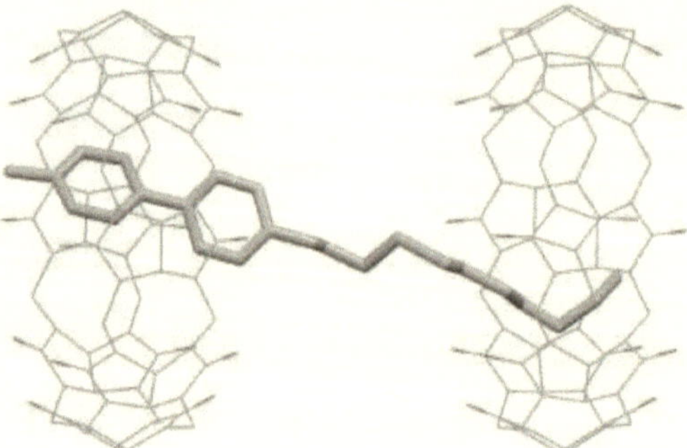

Figure 4.19: Structure of the 1:2 MVC_{11} : CB[8] complex.

Since viologen derivatives with alkyl tails are able to show intercavity folding, this structure with two separate hosts for the guest head and tail is unprecedented. In the crystal structure these 1:2 complexes of MVC_{11} : CB[8] were packed perpendicular to each other. A closer look at the packing shows that the adjacent dumbbell shaped complexes are oriented such that one CB[8] encapsulating a alkyl tail end (shown in green in Figure 4.20) of the guest closes the half of the cavity of the other CB[8] that encapsulates the head of another guest (shown in blue) with C−H···O interactions that are typical of CB[n] packing. This alternate closing of half the cavity leads to a chain arrangement of the complexes as shown in Figure 4.20.

Figure 4.20: Packing of the 1:2 MVC_{11} : CB[8] complex. CB[8] encapsulating the aromatic region of the guest are shown in blue while CB[8] encapsulating the aromatic tail are shown in green. The guest is shown in yellow.

ITC titration data carried out at a concentration of 0.14 mM with respect to the CB[8] in the cell showed an inflection point at a molar ratio around one (N = 1.1) indicating the formation of a 1:1 assembly in aqueous solution (Figure 4.21). The fitting of the observed exothermic curve to a 1:1 binding model gave an association constant $K_a = 1.78 \times 10^7$ M^{-1}. The parameters calculated were ΔH = -55.23 kJ mol^{-1}, ΔG = -41.30 kJ mol^{-1} and TΔS = -13.93 kJ mol^{-1}. The association is driven strongly by enthalpy and less by entropy.

The ^{1}H−NMR spectrum of MVC_{11} upon complexation with CB[8] showed small shifts in the aromatic signals ($H_{a,b,c,d}$) of the viologen. However, all the peaks corresponding to the aliphatic chain were shifted remarkably upfield. Additionally, the peaks were separated from each other. This indicates that the aliphatic tail of the viologen is encapsulated in the hydrophobic cavity of CB[8]. In particular, the protons at the end of the chain close to the double bond H_9 showed a very large upfield shift.

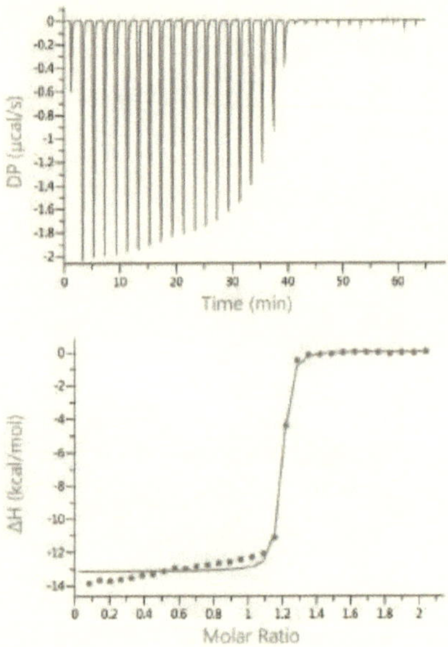

Figure 4.21: Isothermal calorimetric data obtained in the titration of CB[8] in water adding increments of MVC_{11}. [CB[8]]=0.14 mM. Experiments carried out at T $= 25\,°C$

From investigations of MVC_{11} in solution, the unusual 1:2 complexation mode turned out to be peculiar to the solid state. While in solution, studies point to a 1:1 stoichiometry in solution. The majority of 1:1 complexes exhibit enthalpy changes lower than -40 kJ mol^{-1}, which corresponds to the range expected for binary complex formation, where the CB[8] cavity is partially dehydrated.[24] Though the complex shows 1:1 stoichiometry in solution, the thermodynamic data for CB[8] - MVC_{11} complex determined by ITC, shows a relatively high $\Delta H = $ -55.23 kJ mol^{-1}. This value is in line with ternary complex formation where majority of the water molecules are expelled from the cavity. This could be either due to the U-shaped backfolding of the alkyl chain of the molecule, or due to the formation of 2:2 quarternary complexes in solution. The investigation of this behaviour by mass spectroscopy which is currently ongoing is expected to shed further light on this aspect.

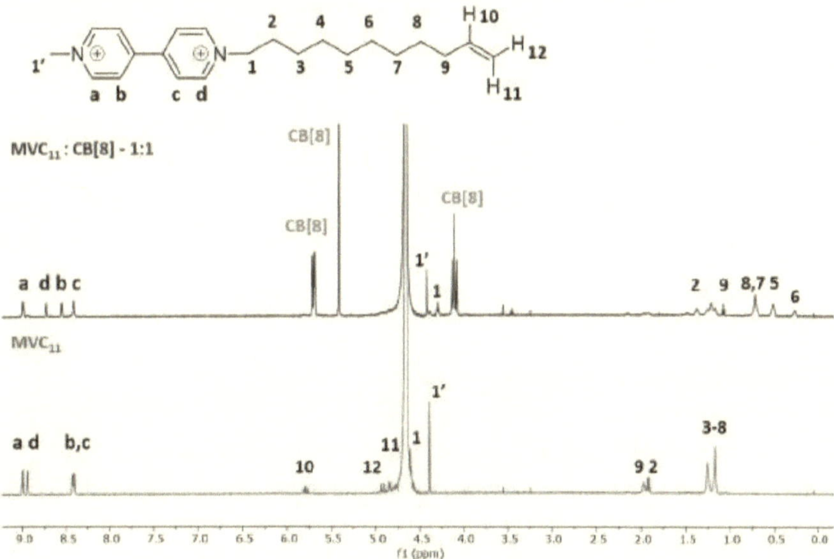

Figure 4.22: ^{1}H–NMR spectrum of MVC_{11} and MVC_{11} : CB[8] 1:1 in D_2O, 600 MHz. $c[MVC_{11}] = 0.2$ mM.

4.5 Complexation of Thiazole Orange with CB[8]

Alongside cyanine dyes, another dye that has gained interest because of its interaction with biomolecules and application in the detection of trace amounts of DNA and RNA is Thiazole Orange (TO) (Figure 4.23). The photophysical behavior of this dye is also remarkably sensitive to its microenvironment for multiple reasons.

TO has the tendency to aggregate and shows virtually no fluorescence in aqueous solution because of the highly efficient nonradioactive coupled photoisomerisation and torsional motion between the benzothiazole and quinoline heterocycles.[11, 25] However, upon interaction with nucleobases or in viscous solutions, it displays over a hundred fold enhancement in its emission. Mohanty and coworkers had reported that the light up of thiazole orange upon CB[8] encapsulation occurs due to 2:2 complex formation between CB[8] and TO as shown in Figure 4.24 by studying spectroscopic changes upon the ad-

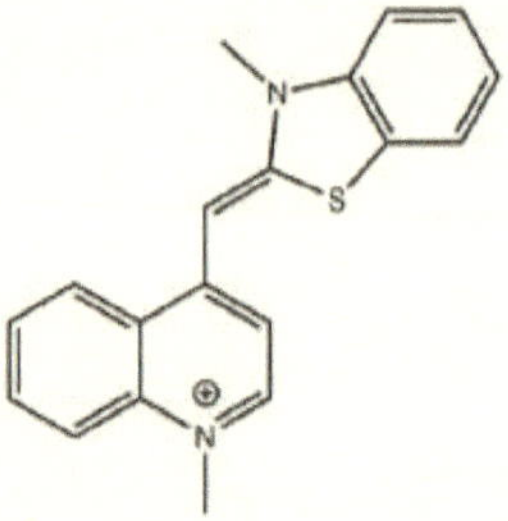

Figure 4.23: Structure of Thiazole Orange (TO).

dition of the host.[26] However no crystal structures of these complexes were reported. They had also gone on to show that the complex responded to certain stimuli such as a competitive guests and metal ions and had also suggested ternary complex formation with other CB[8] guests. We therefore decided to investigate whether TO would form ternary complexes with quencher guests synthesised for use in self diagnostic composites as described in Chapter 2.

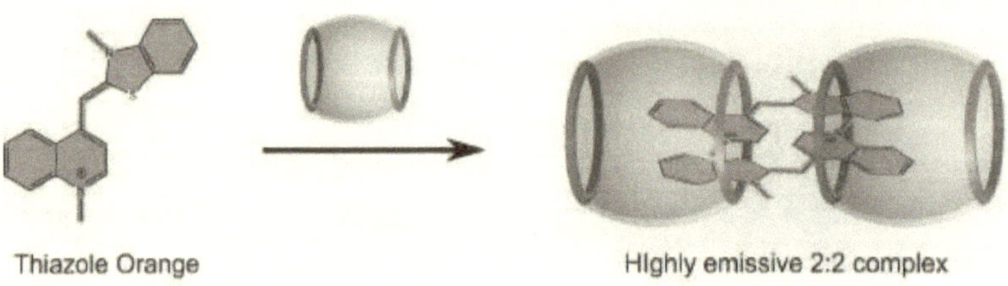

Figure 4.24: Schematic representation of 2:2 complex formation causing fluorescence turn on.

4.5.1 Spectroscopic changes on encapsulation with CB[8]

Initial experiments were carried out by titrating a 1 mM aqueous solution of TO with CB[8]. The spectrum of TO in water an absorption maxima at 500 nm with a shoulder at 480 nm and a small peak at 430 nm which corresponds to higher H aggregates (Figure 4.25). The peak at 470 nm is attributed to that of the H-dimer.[25] Upon addition of CB[8] there was a marked decrease in the peak at 500 nm followed by the appearance of a new peak at 490 nm and the appearance of a shoulder at 525 nm. These are in correspondence with literature values and are caused by the stabilisation of a 2:2 complex of CB[8]:TO as shown in Figure 4.24.

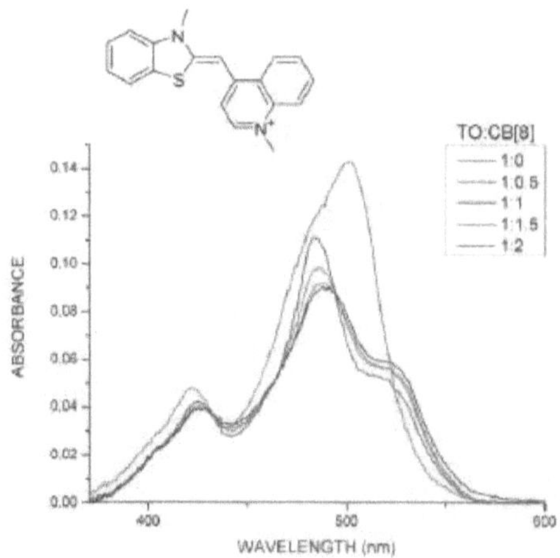

Figure 4.25: Changes in UV-Vis spectra of TO upon titration with CB[8], 1 μM.

This 2:2 complex formation was also supported by the changes in the fluorescence spectrum where addition of CB[8] caused a huge increase in the fluorescence intensity, with the solution of free thiazole orange showing no fluorescence before addition of CB[8] (Figure 4.26). Due to the large size of the CB[8] the ditopic dye forms a 2:2 complex with 2 molecules of CB[8]. Since in the 2:2 complex the excited state torsional motion and photo-isomerisation are very effectively hindered, it follows that the relaxation of the excited state by non radiative pathways is reduced and the intensity of fluorescence is increased. This is similar to the case of the Cy3 dye, however the contrary effect occurs.

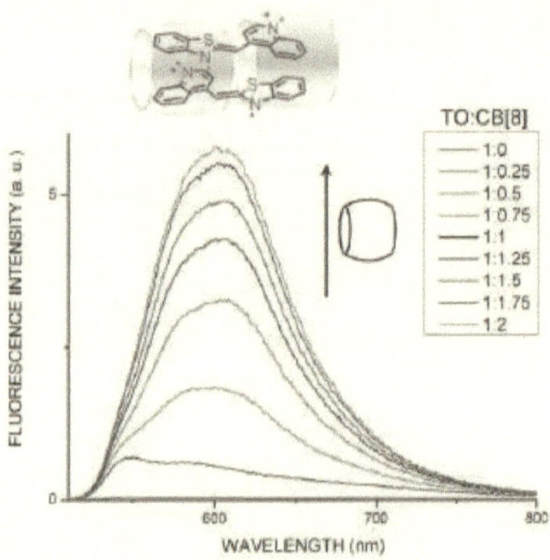

Figure 4.26: Changes in fluorescence spectra of TO upon titration with CB[8], 1 µM.

Conversion of 2:2 to 1:1:1 heteroternary complex

Since 2:2 complexes of CB[8] and TO have been established, the effect of addition of a third guest to the system was investigated. It has already been shown that the addition of adamantyl amine (at pH = 4.75) which forms a 1:1 complex with CB[8], switches off the fluorescence by kicking out the all the thiazole orange from within the CB[8] cavity. Azobenzene (AZO) and dibenzofuran (DBZ) derivatives have been used earlier in this book to cause effective quenching of a 1:1 perylenediimide-CB[8] complex. These molecules are known to be effective second guests with CB[8] and have a low binding affinity with CB[8] without the presence of this first guest. Upon titration of a 1:1 solution of CB[8] : TO with a solution of AZO (Figure 4.27) and a solution of DBZ (Figure 4.28), upon addition of an excess of the quencher the fluorescence was observed to decrease in both cases. Additionally, the fluores-cence band was slightly red shifted from the peak of the monomer. This points towards ternary complex formation, wherein the charge transfer between the TO and the second guest quenches the fluorescence emission. DBF is more effective than AZO in forming the heteroternary complex.

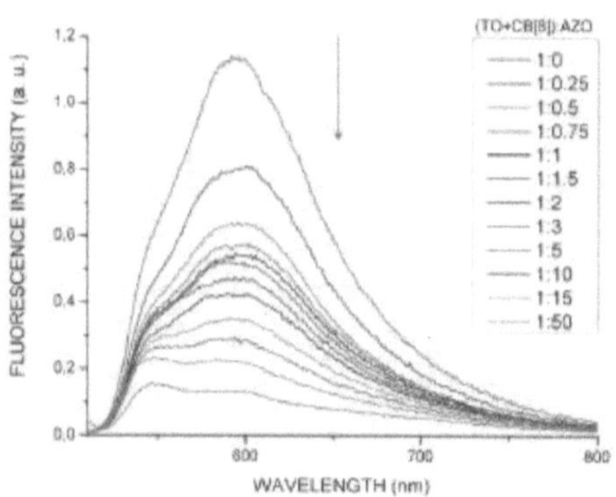

Figure 4.27: Changes in Fluorescence spectra of TO:CB upon addition of AZO 1 µM.

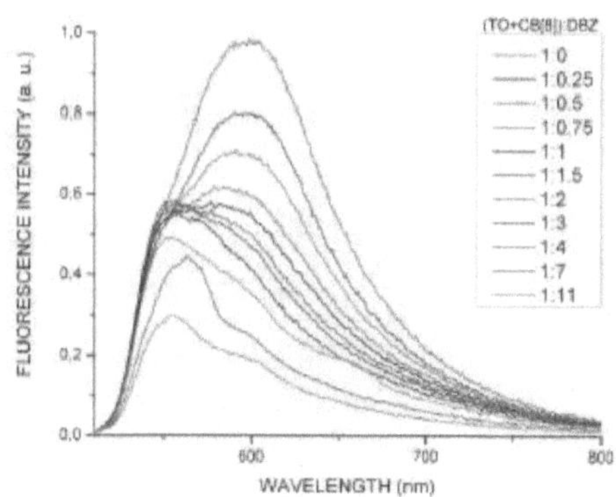

Figure 4.28: Changes in Fluorescence spectra of TO:CB upon addition of DBZ guest, 1 µM.

4.5.2   Solid state interaction with CB[8]

While at low concentrations, the 2:2 complex between CB[8] and TO is preferentially formed, Pang and coworkers suggested that at higher concentrations

TO forms linear supramolecular polymers with CB[8].[27] We attempted to grow crystals from solutions of CB[8] and TO in order to reveal the packing of complex in the solid state. However, it was observed that repeatedly, needle shaped yellow crystals were formed that were determined by single crystal X-ray diffraction to be an inclusion complex of CB[8] and a 3-methyl-1,3-benzothiazol-3-ium cation. This was unexpected and is thought to originate from the cleavage of TO under the acidic medium of the CB[8] solution. The unit cell is unusually large, and the asymmetric unit is shown in Figure 4.29, and consists of one whole CB[8] (A), two half CB[8] (B and C) macrocyles, one 3-methyl-1,3-benzothiazol-3-ium tosylate, and 42 lattice water molecules, of which 18 were treated with the SQUEEZE procedure (see the experimental section for details)

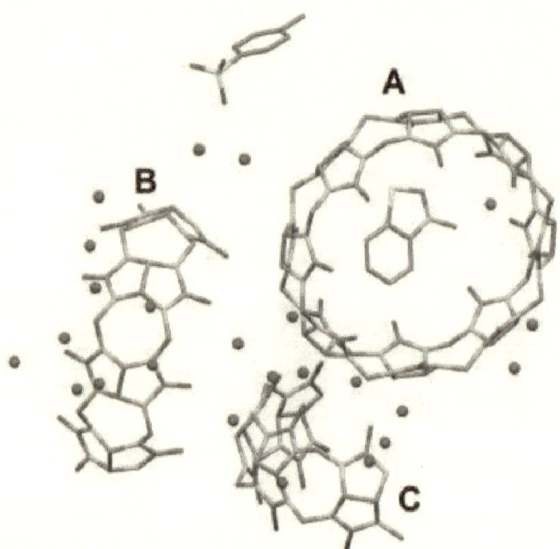

Figure 4.29: Molecular structure of the asymmetric unit of the inclusion complex. The oxygen atoms of the water molecules are represented as red spheres of arbitrary radius.

The main feature of the structure is the inclusion compound formed by CB[8]A and the 3-methyl-1,3-benzathiazol-3-ium cation (see Figure 4.30 (L)) at an almost horizontal angle. The guest is held inside the host via weak

C–H···O bonds between the hydrogen atoms of the cation and the oxygen atoms at the upper and lower rim of the macrocycle. The perfect fit in terms of dimensions of the cation and the cavity of CB[8] is highlighted in the space filling mode of Figure 4.30 (R). The packing in the crystal structure is driven by the extensive network of H bonds involving the macrocycles, the water molecules and the tosylate anion; the relative positions of the CB[8]s in the unit cell is shown in Figure 4.31.

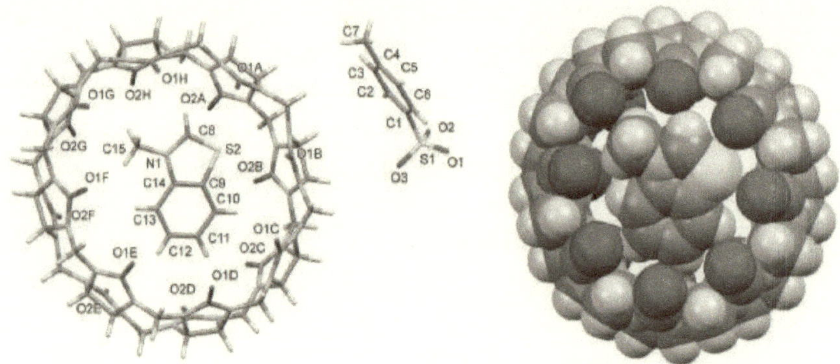

Figure 4.30: Molecular structure of CB[8] A and 3-methyl-1,3-benzothiazol-3-ium tosylate with partial labelling scheme. Water molecules have been omitted for clarity. Right: space filling mode of the host-guest complex formed by the macrocycle and the cation.

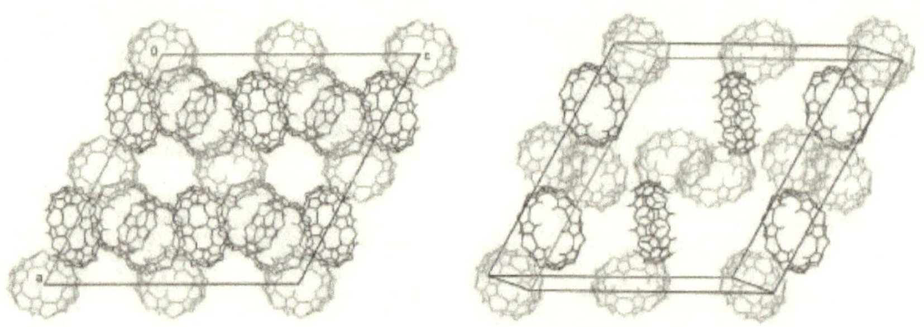

Figure 4.31: Left: packing of the CB[8] macrocycles viewed along the *b*-axis direction of the unit cell. Colour code: CB[8] A in red, CB[8] B in green and CB[8] C in blue. Right: details of the reciprocal orientations of CB[8] B and C (right). H atoms and lattice water molecules have been omitted for clarity

4.6 Ternary complexes of Azine and Thiazine dyes in CB[8]

Azine, oxazine and thiazines, being among the earliest dyes to be discovered, are a historically important class of dyes. At present, their use as synthetic dyes is limited, but many members of this dye family have found wide uses in other areas of research. These dyes have a planar heteroaromatic molecular structure with an inherent positive charge, have very high absorption coefficients and show fairly intense fluorescence. They also tend to show high photoconductivity and undergo reversible oxidation processes, and are therefore widely used as dyes, antioxidants, and as stains for microscopy in biological systems.

The interaction of two dyes of this family with CB[8] was investigated. azure a (AA) and neutral red (NR) are dyes with the same basic dye skeleton as shown in Figure 4.32(a).

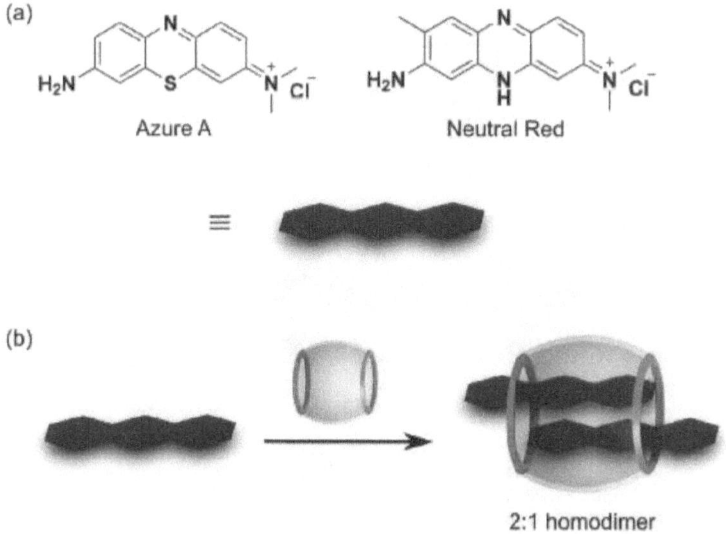

Figure 4.32: (a) Structure of dyes investigated (b)Schematic representation of formation of homodimers of AA and NR in the presence of CB[8].

At high concentrations, these dyes tend to form aggregates in aqueous medium leading to fluorescence quenching, but are reasonably fluorescent at low concentrations. Due to their intrinsic monocationic charge, these dyes can form homodimers in CB[8]. Encapsulation of two guest molecules by one CB[8] is expected to cause a decrease in fluorescence intensity due to quenching within the dimer as shown in Figure 4.32(a). Turn on of fluorescence can therefore be triggered by a competitive binder.[28]

4.6.1   Azure A

To avoid the self aggregation of the dye, studies were carried out at a low concentration of 5 μM where the absorption spectrum showed the presence of the monomeric species. The UV Vis spectrum of Azure A in water shows a peak at 630 nm. When progressively increasing amounts of CB[8] were added, there was a sharp decrease in the peak at 630 nm. This was followed by the appearance of a new blue shifted peak at 595 nm. Upon the addition of 0.5 equivalents of CB[8] the monomeric peak had completely disappeared (Figure 4.33).

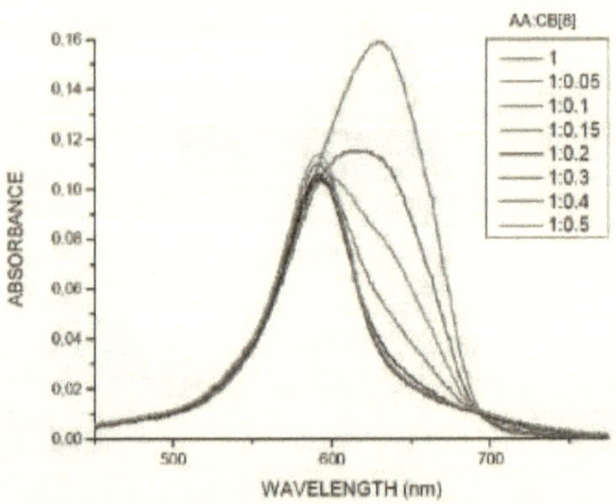

Figure 4.33: Changes in UV-Vis spectra of AA upon titration with CB[8], 5 μM

As expected, the fluorescence intensity of the AA dye decreased almost completely upon addition of 0.5 equivalents of CB[8], as shown in Figure 4.34

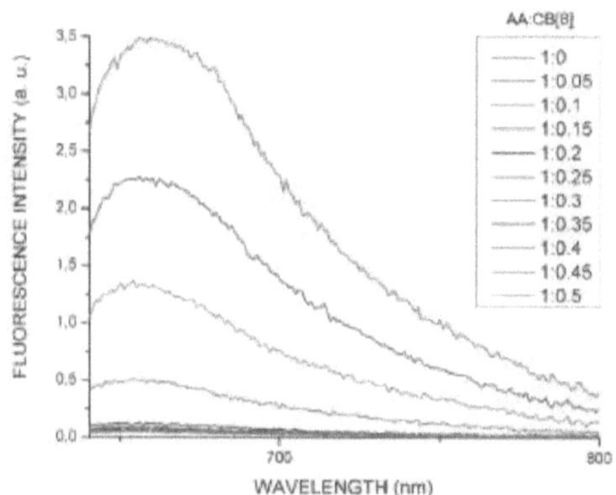

Figure 4.34: Changes in fluorescence spectra of AA upon titration with CB[8], 5 µM

also pointing to the formation of a dimer within the cavity leading to loss of fluorescence. Addition of methyl viologen or adamantyl amine to the system results in an immediate reappearance of the fluorescence of the monomer due to displacement of the dimer from the cavity.

4.6.2 Neutral Red

The behaviour of NR upon addition of CB[8] was very similar to that of AA. NR in aqueous solution has an absorption maxima at 560 nm. (Figure 4.35) This peak disappears and is followed by the appearance of a blue shifted peak at 500 nm. Similarly, the fluorescence intensity of the dye decreases with addition the host (Figure 4.36) as a result of the quenching of the dye pair within the cavity.

All attempts to grow single crystals from solutions of these complexes under a wide variety of experimental conditions failed. AA and NR are well soluble in water and increase the solubility of CB[8] drastically upon complexation. It was also interesting to note is that these complexes could be formed by solvent assisted grinding of the dye molecules with CB[8].

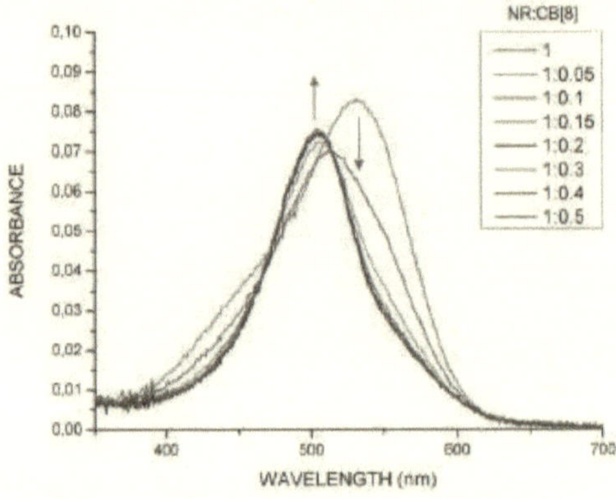

Figure 4.35: Changes in UV-Vis spectra of NR upon titration with CB[8], 5 µM

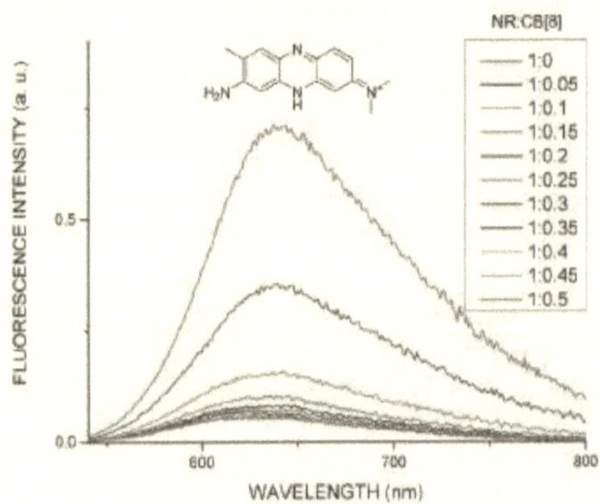

Figure 4.36: Changes in fluorescence spectra of NR upon titration with CB[8], 5 µM

4.7 Conclusions

In conclusion, we present a series of CB[8] host guest complexes spanning different stoichiometries from 2:1, 1:1, 2:2 and finally 1:2 with respect to the CB[8] host molecule including including novel crystal structures of a 1:1 and

2:1 and 1:2 cucurbituril complex. Additionally, the effect of encapsulation of fluorescent dyes by cucurbiturils was studied. Some of these complexes, especially of dyes that show fluorescence in the near IR region, also show promise for the preparation of damage reporting materials with higher depth penetration allowing the detection of cracks deeper within the material. Dye encapsulation by CB can also help in the design of displacement assays, in which the fluorescence regeneration upon analyte binding is greatly enhanced.

Crystal structures and solution experiments often lead to different complexation outcomes, as already reported in Chapter 3. These discrepancies are mainly due to the necessity of additional extracavity interactions to aid the crystallisation process which alters the intrinsic binding ability of CB[8] in water.

4.8   Experimental section

Synthesis

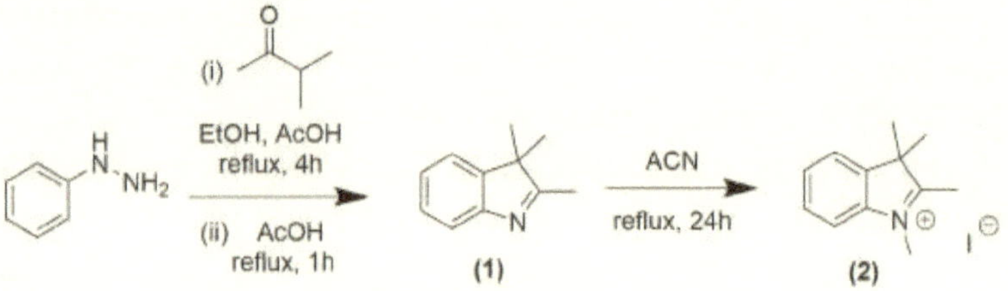

Scheme 4.2: Synthetic scheme for Cy3 precursor

Synthesis of 2,3,3-trimethyl-3H-indole (1)

Phenylhydrazine (9.82 mL, 0.1 mol) and 3-methyl-2-butanone (12 mL, 0.11 mol) were dissolved in a mixture of anhydrous EtOH (30 mL) and CH_3COOH (three drops). The solution was refluxed for 4 h, then allowed to cool to room temperature, and the solvent was removed under reduced pressure. To the resulting viscous residue was added CH_3COOH (30 mL), and the solution was refluxed for 1 h under Ar atmosphere, then allowed to cool to room temperature. Most acidic acid was removed under reduced pressure. The resulting residue was washed with saturated aqueous solution of $NaHCO_3$ to neutralize the pH of the solution. The resulting mixture was extracted with DCM/H_2O. The organic solution was dried with anhydrous Na_2SO_4 and concentrated to provide the crude product, which was purified by column chromatography (silica gel, hexane/EtOAc, 85:15, v/v) to give compound **1** (12.7 g) as light yellow oil, which changed to red with time. Yield, 80%.

^{1}H NMR (400 MHz, $CDCl_3$): δ 7.55 (d, J = 8.0 Hz, 1H), 7.26 7.29 (m, 2H), 7.20 (dd, J1 = 4.0 Hz, J2 = 8.0 Hz, 1H), 2.28 (s, 3H), 1.29 (s, 6H);

^{13}C NMR (100 MHz, $CDCl_3$): δ 15.18, 23.01, 53.54, 119.79, 121.26, 125.16, 127.57, 145.52, 153.32, 188.11;

ESI-MS: Calcd. For $C_{11}H_{13}N$ 159.1048; Found, 159.1.

Synthesis of 1,2,3,3-tetramethyl-3H-indol-1-ium iodide (2)

Compound **1** (4.773 g, 30 mmol) was dissolved in a mixture of methyl iodide (24.11 mL, 300 mmol) and acetonitrile (60 mL) under Ar. The solution was refluxed for 24 h, then allowed to cool to room temperature, and the resulting solid was collected by filtration. The solid was washed with acetone and recrystallised with EtOH to give compound **2** as yellow crystalline compound (4.246 g). Yield, 45%.

^{1}H NMR (300 MHz, DMSO–d$_6$): δ 7.93-7.90 (m, 1H), 7.85-7.82 (m, 1H), 7.64-7.61 (m, 2H), 3.98 (s, 3H), 2.78 (s, 3H), 1.53 (s, 6H).

^{13}C NMR (300 MHz, DMSO–d$_6$): 196.4, 142.6, 142.1, 129.8, 129.3, 123.8, 115.6, 54.5, 35.3, 22.2, 14.7

ESI-MS: C$_{12}$H$_{15}$N [M+H]$^+$ m/z: found 174.25, calc. 174.10.

Synthesis of Cy3

Scheme 4.3: Synthetic scheme for Cy3

2,3,3-trimethyl-3H-indole (**2**) (0.63 g, 4 mmol) was refluxed with triethylorthoformate (0.6 mL, 8 mmol) in 3 mL of pyridine at 120 °C for 1 h. The reaction mixture was then cooled to room temperature and excess diethylether was carefully added to the solution to precipitate the product. The crude dye was recrystallised from ethanol to give the product with a 58% yield.

^{1}H NMR (300 MHz, DMSO–d$_6$): δ 8.34 (t, J=13.6 Hz, 1H), 7.64 (d, J=7.4 Hz, 2H), 7.46 (m, 4H), 7.29 (m, 2H), 6.45 (d, J=13.6 Hz), 3.65 (s, 6H), 1.69 (s, 12H).

^{13}C NMR (300 MHz, DMSO–d$_6$): δ 174.8, 149.9, 143.3, 141.0, 129.1, 125.8, 123.0, 112.1, 103.2, 49.3, 32.0, 27.7.

X-ray crystallographic section for AzPy - CB[8]

Synthesis of AzoPy-CB[8] crystals

Azopyridine (1.842 mg, 0.01 mmol), CB[8] (9 mg, 0.0005 mmol) were taken in a 5 mL vial with 2 mL of milliQ water. The vial was capped and kept at 60 °C until the components dissolved. On slow evaporation at room temperature, red hexagonal crystals were observed in 4 weeks.

X-ray structure determination

The crystal structures of the compound was determined by X-ray diffraction methods. Crystal data and experimental details for data collection and structure refinement are reported in Table 4.1. Intensity data and cell parameters were recorded at 200(2) K on a Bruker D8 Venture Photon II diffractometer equipped with a CCD area detector, using the CuKa radiation ($\lambda = 1.54178$). The raw frame data were processed using the programs SAINT and SADABS. The ORTEP view is shown in Figure 4.37

The structure was solved by Direct Methods using the SIR97 program B and refined on F_o^2 by full-matrix least-squares procedures, using the SHELXL-2014/7 program in the WinGX suite v.2014.1.[29] All non-hydrogen atoms were refined with anisotropic atomic displacements. The carbon-bound H atoms were placed in calculated positions and refined isotropically using a riding model with C-H ranging from 0.95 to 0.99 Å and U_{iso}(H) set to $1.2U_{eq}$(C). The O-bound H atoms were found in the difference Fourier map and then refined setting the O-H distance to 0.87 Å and U_{iso}(H) to $1.5U_{eq}$(O). The weighting schemes used in the last cycle of refinement was $w = 1/[\sigma^2 F_o^2 + (0.3540P)^2]$ where $P = (F_o^2 + 2F_c^2)/3$. One of the pyridine rings of the 4,4'-azopyridine ligand was found disordered over two equivalent positions each with an occupancy factor of 0.5.

An extensive network of H bonds involving the water molecules and the nitrate counter ions helps consolidate the crystal structure and connects the chains with each other. The overall arrangement in space of the CB[8] macro-cycles resulting from these multiple interactions is shown in Figure 4.38.

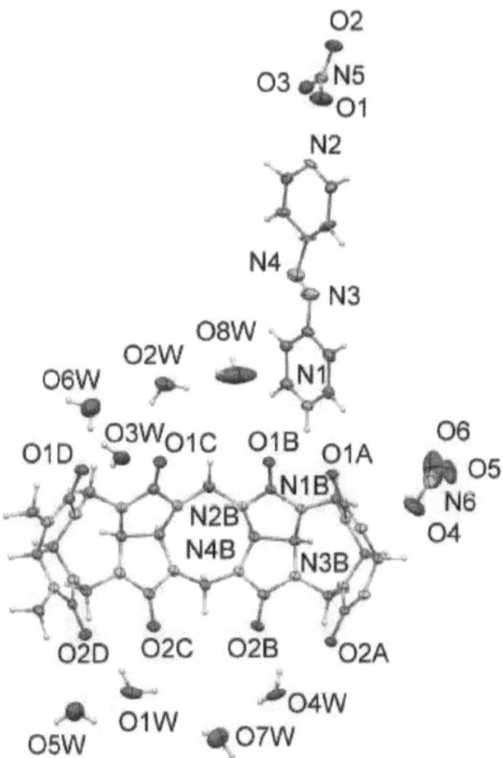

Figure 4.37: ORTEP view of the asymmetric unit of AzPy-CB[8] complex with partial labelling scheme and ellipsoids drawn at the 20% level. The disordered atoms have been omitted for clarity.

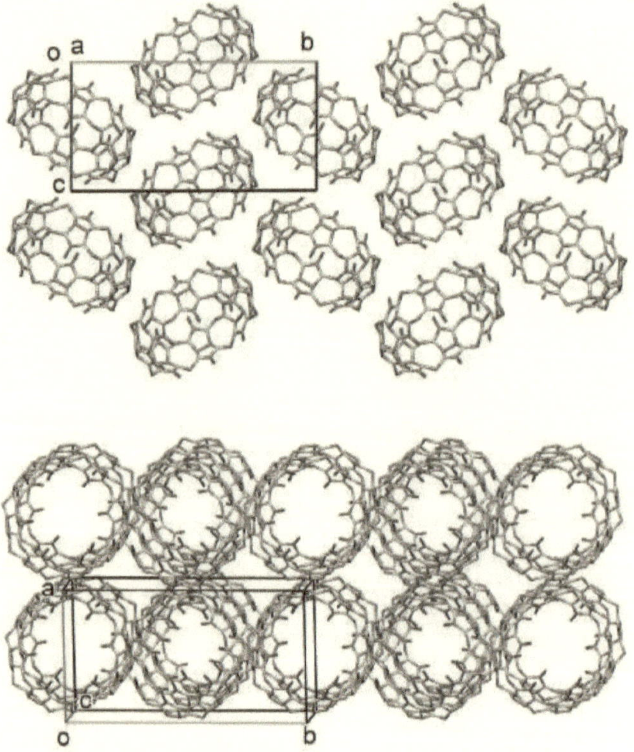

Figure 4.38: Packing of the CB[8] macrocycles viewed along the a- and c-axis direction of the unit cell (above and below, respectively).

Parameter	Value
Empirical formula	$C_{68}H_{83}N_{43}O_{33}$
Formula weight	2030.77
Crystal system	Monoclinic
Space Group	$P2_1/c$
a/Å	14.4470(10)
b/Å	25.345(2)
c/Å	13.4836(8)
β/deg	105.557(4)
V/Å^3	4696(4)
Z	2
T/K	200(2)
ρ/cm^3	1.418
μ/mm^{-1}	0.994
F(000)	2112
total reflections	76497
unique reflections (R_{int})	8101 (0.0596)
observed reflections $[F_o/ > 4/\Sigma(F_o)]$	6409
GOF on F^{2a}	0.998
R indices $[F_o > 4\sigma(F_o)]^b R_1, wR_2$	0.1034, 0.3308
largest diff. peak and hole eÅ^{-3}	1.416, -0.540

[a]Goodness-of-fit $S = [\Sigma(F_o^2 - Fc2)2/(n - p)]1/2$, where n is the number of reflections and p the number of parameters. [b]$R1 = \Sigma||F_o| - |F_c||/\Sigma|F_o|, wR_2 = [\Sigma[w(F_o^2 - F_c^2)^2]/\Sigma[w(F_o^2)^2]]^{1/2}$.

Table 4.1: Crystallographic Data for CB[8]-AzPy

Appendix B

B.1   X-ray crystallographic data for Ligand Cy3

The crystal structure of compound Cy3 was determined via X-ray diffraction data on single crystals obtained by slow evaporation of an aqueous solution of 0.1 mM concentration. The asymmetric unit (see Figure B.1) consists of two independent molecules (A and B), each of general formula $(C_{25}H_{29}N_2)^+I^-$, and a lattice water molecule. Both A and B are essentially planar, with the exception of the methyl groups (C12A, C13A, C24A, C25A, C12B, C13B, C24B and C25B) on the 1-methylindoline and 1-methylindolinium fragments, which project out of the mean plane due to the sp^3 hybridisation of the carbon atoms C4A, C16A, C4B and C16B. The two molecules show a high degree of electronic localisation, as demonstrated by the relevant bond distances (see Table B.1). Figure B.2 shows an overlap of molecule A (red) and B (blue), highlighting their very similar conformation. In the lattice, the main interactions responsible for the packing are C−H···O and dispersion forces.

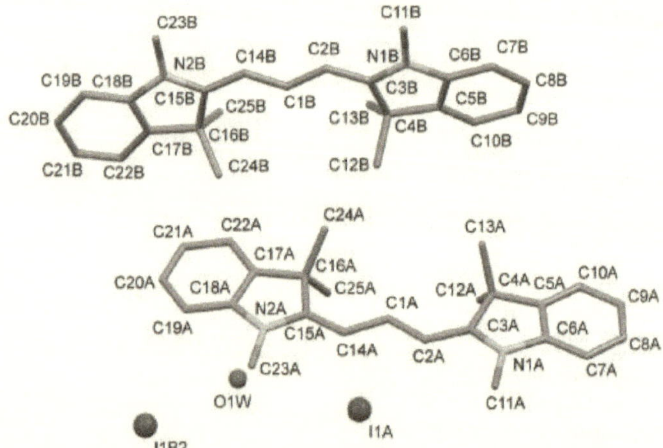

Figure B.1: Molecular structure of the asymmetric unit of Cy3 with partial labelling scheme. Hydrogen atoms have been omitted for clarity.

The crystal structures of compound Cy3 was determined by X-ray diffrac-

N1A-C3A	1.352(5)	N1B-C3B	1.356(5)
N1A-C6A	1.397(5)	N1B-C6B	1.401(7)
C2A-C3A	1.389(5)	C2B-C3B	1.3767(7)
C1A-C2A	1.389(5)	C1B-C2B	1.413(6)
C1A-C14A	1.376(5)	C1B-C14B	1.350(8)
C14A-C15A	1.396(5)	C14B-C15B	1.390(7)
N2A-C15A	1.347(5)	N1B-C15B	1.341(7)
N2A-C18A	1.415(5)	N2B-C18B	1.423(7)

Table B.1: Selected bond distances for Cy3 ligand

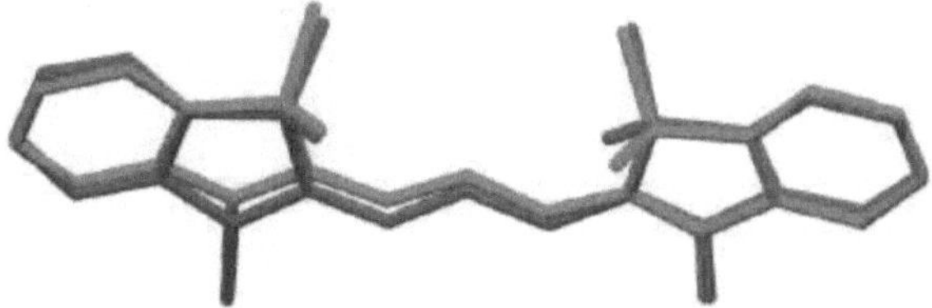

Figure B.2: Overlap of the independent molecule A (red) and B (blue).

tion methods. Crystal data and experimental details for data collection and structure refinement are reported in Table B.2. Intensity data and cell parameters were recorded at 200(2) K on a Bruker D8 Venture Photon-II diffractometer equipped with a CCD area detector, using the CuKa radiation (λ = 1.54178). The raw frame data were processed using the programs SAINT and SADABS. The structure was solved by Direct Methods using the SIR97 program and refined on Fo2 by full-matrix least-squares procedures, using the SHELXL-2014/7 program in the WinGX suite v.2014.1. All non-hydrogen atoms were refined with anisotropic atomic displacements. The carbon-bound H atoms were placed in calculated positions and refined isotropically using a riding model with C-H ranging from 0.95 to 0.97 Å and Uiso(H) set to 1.2Ueq(C) or 1.5Ueq(C) in case of the methyl groups. The H atoms of the water molecule could not be located in the difference Fourier map. One of the two iodide anions was found disordered over two equivalent positions each with an occupancy factor of 0.5. The weighting scheme used in the last cycle of refinement was $w = 1/[\sigma^2 F_o^2 + (0.1351P)^2 + 5.2161P]$ where $P = (F_o^2 + 2F_c^2)/3$.

Parameter	Value
Empirical formula	$C_{25}H_{29}IN_2 \ \frac{1}{2}H_2O$
Formula weight	493.41
Crystal system	Orthorhombic
Space Group	*Pbcn*
a/Å	27.237(3)
b/Å	14.716(2)
c/Å	24.347(3)
V/Å^3	9759(2)
Z	16
T/K	200(2)
ρ/cm^{-1}	1.343
μ/mm^{-1}	10.406
F(000)	4016
total reflections	159030
unique reflections (R_{int})	8366 (0.0557)
observed reflections $[F_o/ > 4/\Sigma(F_o)]$	7306
GOF on F^{2a}	0.998
R indices $[F_o > 4\sigma(F_o)]^b R_1, wR_2$	0.0526, 0.1853
largest diff. peak and hole eÅ^{-3}	2.052, -0.667

[a]Goodness-of-fit $S = [\Sigma(F_o^2 - Fc2)2/(n - p)]1/2$, where n is the number of reflections and p the number of parameters. [b]$R1 = \Sigma||F_o| - |F_c||/\Sigma|F_o|, wR_2 = [\Sigma[w(F_o^2 - F_c^2)^2]/\Sigma[w(F_o^2)^2]]^{1/2}$.

Table B.2: Crystallographic Data for Cy3 ligand

Epilogue

Smart materials that can sense and react to external stimuli constructed by exploiting supramolecular assemblies have recently received significant attention due to their potential applications in a number of fields such as self-healing materials and sensorsHost-guest inclusion complexes formed by macrocyclic hosts such as cyclodextrines, calixarenes and cucurbiturils providing an exciting platform by participating in dynamic but strong interactions with guest molecules. Of these hosts, cucurbit[n]urils are particularly interesting and can participate in the formation of various supramolecular assemblies having unprecedented properties through its distinctive molecular recognition ability and the high selectivity of its different homologues to various guests, all of which plays a very important role in the formation of CB[8] based functional supramolecular materials.

This book describes the inclusion chemistry of the well-known macrocyclic compound cucurbit[8]uril, one of the larger homologues of the CB[n] family, focusing on the encapsulation of fluorescent dyes and shows an interesting application of these complexes by the design of supramolecular probes based on CB[8] ternary complexes to identify early stage damage in carbon fibre reinforced composite materials.

Carbon fiber reinforced composites have emerged as the materials of choice for a wide variety of structural applications over the last few decades. Their unique combination of light weight and high mechanical strength is a huge advantage for load bearing applications. Their Achilles' heel, however, is sudden catastrophic failure, induced by fatigue and microscopic cracks. The ability to anticipate such failure by detecting early signs of irreversible damage and in a structural element is a major issue in advancing the use of carbon fiber composites.

Chapter 1 addresses this issue by the introducing a supramolecular probe formed by a fluorescent probe and quencher encapsulated by a CB[8] host molecule into the epoxy curing agent of the composite as a supramolecular cross-linking agent. The cross-linked CB[8] ternary complex provides a reliable fluorescence response upon mechanically driven de-complexation induced by strain and fatigue. Since the interactions holding together the host-guest complex are weaker than the polymer covalent bonds, fluorescence emission can be easily obtained at low strain.

A combination of uniaxial tensile and compression testing of carbon fiber epoxy composite specimens followed by fluorescence microscopy enabled the detection of strained regions and identification of sites for crack nucleation and growth in a non-destructive manner. Even more compelling was the ability to report damage by fatigue, the major source of concern in the real-world use of such materials. A major advantage of this method is that the low amount of additive (0.003 w % in the final composite material) which results in preservation of the mechanical characteristics of the matrix. Moreover, the introduction of the reporting agent as additive in the epoxy formulation does not alter the composite structural elements' fabrication.

In order to get a better fundamental understanding of the self assembly of cucurbit[8]uril especially in the presence and absence of extraneous agents we studied the solid-state inclusion chemistry of four complexes under different conditions. Here the attempt was to produce porous crystalline materials based on cucurbit[8]uril building blocks. While we successfully obtained four crystal structures, our attempts to further access the porosity of these structures was hampered by crystallographic limitations due to poor quality crystals formed by the cucurbituril macrocycle. Nevertheless, the most promising of these structures, the study of the tubular framework of CB[8] synthesised through direct metal coordination of Cu^{2+} is currently under further investigation.

When considering dye molecules for damage reporting materials, there is an interest to make these sensing systems more versatile. One of the ways to achieve this is to move to near IR dyes due to their improved depth-penetration of light. In the final chapter, solid-state and solution-phase encapsulation of additional dye molecules of interest in these systems by CB[8] was studied. The interaction of CB[8] and a model cyanine dye was explored and the formation of a supramolecular polymer in the solid state was characterised. Additionally, three novel crystal structures of CB[8] with guest molecules were reported, going from a 2:1, 1:1 and an unusual 1:2 stoichiometry with respect to the host.

In summary, we have shown that cucurbit[8]uril based ternary complexes based on fluorescent guest molecules form a very good system for measuring strain in composite materials. These results have been patented in view of their technical potential. However in order to improve these sensors and make such systems more versatile for application in other materials, there are two aspects that need to be addressed. First is to improve the fundamental understanding of the self assembly processes of CB[8] in the presence of extraneous materials and second, to investigate additional dyes in the near IR range. The work reported in the final part of this book contributes significantly in this direction.

Outreach

Peer Reviewed Journal Publications

- **A. D. Das**, G. Mannoni, A. Fruh, D. Orsi, R. Pinalli and E. Dalcanale, *ACS Appl. Polym. Mater.*, **2019**, 1, 2990–2997

Patent Applications

- E. Dalcanale, **A. D. Das**, R. Pinalli, P. Gherardi; Self-diagnostic epoxy resins. EPA 19156705.6 filed 12.02.2019.

Conferences

- 5th Conference on Methods and Applications in Fluorescence, September 2017 at Bruges, Belgium.

- 7th EuCheMS Chemistry Congress: Molecular frontiers and global challenges, August 2018 at Liverpool, UK. Royal Society of Chemistry Best Poster award sponsored by Polymer Chemistry.

- 14th International Symposium on Macrocyclic and Supramolecular Chemistry, June 2019 at Lecce, Italy. Royal Society of Chemistry Best Poster award sponsored by ChemComm.